Machine Learning

Algorithms and Applications

Machine Learning

Algorithms and Applications

Mohssen Mohammed
Muhammad Badruddin Khan
Eihab Bashier Mohammed Bashier

CRC Press
Taylor & Francis Group
Boca Raton   London   New York

CRC Press is an imprint of the
Taylor & Francis Group, an **informa** business

CRC Press
Taylor & Francis Group
6000 Broken Sound Parkway NW, Suite 300
Boca Raton, FL 33487-2742

International Standard Book Number-13: 978-1-4987-0538-7 (Hardback)

Library of Congress Cataloging-in-Publication Data

Names: Mohammed, Mohssen, 1982- author. | Khan, Muhammad Badruddin, author. | Bashier, Eihab Bashier Mohammed, author.
Title: Machine learning : algorithms and applications / Mohssen Mohammed, Muhammad Badruddin Khan, and Eihab Bashier Mohammed Bashier.
Description: Boca Raton : CRC Press, 2017. | Includes bibliographical references and index.
Identifiers: LCCN 2016015290 | ISBN 9781498705387 (hardcover : alk. paper)
Subjects: LCSH: Machine learning. | Computer algorithms.
Classification: LCC Q325.5 .M63 2017 | DDC 006.3/12--dc23
LC record available at https://lccn.loc.gov/2016015290

Visit the Taylor & Francis Web site at
http://www.taylorandfrancis.com

and the CRC Press Web site at
http://www.crcpress.com

To our parents, families, brothers and sisters, and
to our students, we dedicate this book.

Contents

Preface

If you are new to machine learning and you do not know which book to start from, then the answer is this book. If you know some of the theories in machine learning, but you do not know how to write your own algorithms, then again you should start from this book.

This book focuses on the supervised and unsupervised machine learning methods. The main objective of this book is to introduce these methods in a simple and practical way, so that they can be understood even by beginners to get benefit from them.

In each chapter, we discuss the algorithms through which the chapter methods work, and implement the algorithms in MATLAB®. We chose MATLAB to be the main programming language of the book because it is simple and widely used among scientists; at the same time, it supports the machine learning methods through its statistics toolbox.

The book consists of 12 chapters, divided into two sections:

  I: Supervised Learning Algorithms
 II: Unsupervised Learning Algorithms

In the first section, we discuss the decision trees, rule-based classifiers, naïve Bayes classification, k-nearest neighbors, neural networks, linear discriminant analysis, and support vector machines.

In the second section, we discuss the k-means, Gaussian mixture model, hidden Markov model, and principal component analysis in the context of dimensionality reduction.

We have written the chapters in such a way that all are independent of one another. That means the reader can start from any chapter and understand it easily.

MATLAB® is a registered trademark of The MathWorks, Inc. For product information, please contact:

The MathWorks, Inc.
3 Apple Hill Drive
Natick, MA 01760-2098 USA
Tel: 508-647-7000
Fax: 508-647-7001
E-mail: info@mathworks.com
Web: www.mathworks.com

Acknowledgments

We are deeply thankful to all those who have contributed directly or indirectly to the publication of this book. Special thanks go to Dr. Mohsin Hashim, University of Khartoum, Sudan, for his valuable advice.

We would like to thank our colleagues at the Imam Muhammad bin Saud University, Qatar University, and University of Khartoum for their suggestions and encouragement.

We are grateful to Richard O'Hanley of Taylor & Francis Group for his guidance during the preparation of this book.

We would also like to thank all the teams from Taylor & Francis Group/CRC Press for their help in the development and editing of this book.

Authors

Mohssen Mohammed earned a BSc (Honors) in computer science at Future University, Khartoum, Sudan, in 2003. In 2006, he earned an MSc degree in computer science at the Faculty of Mathematical Sciences, University of Khartoum, Sudan. In 2012, he earned a PhD in network security from the Electrical Engineering Department, Cape Town University, South Africa. His PhD dissertation was titled "Automated Signature Generation for Zero-Day Polymorphic Worms Using a Double-Honeynet." His areas of interest include network and information security with a focus on malware detection and analysis methods. Dr. Mohammed has published more than 15 papers at international conferences and in journals. His first book, *Automatic Defense against Zero-Day Polymorphic Worms in Communication Networks*, was classified by IEEE as one of the best books in network security. He is an associate professor at the College of Computer Studies and Information Technology, University of Garden City, Khartoum, Sudan. From December 2012 to June 2016, he was an assistant professor at the College of Computer and Information Sciences, Al-Imam Muhammad Ibn Saud Islamic University, Riyadh, Saudi Arabia.

Muhammad Badruddin Khan earned a PhD in 2011 at the Tokyo Institute of Technology, Japan. Since 2012, he is a full-time assistant professor in the Department of Information Systems of Al-Imam Muhammad Ibn Saud Islamic University.

His research interests are mainly focused on data and text mining. He is currently involved in a number of research projects related to machine learning and Arabic language including Arabic sentiment analysis, improvement of Arabic semantic resources, intelligent Arabic search engine, stylometry, Arabic chatbots, trend analysis using Arabic Wikipedia, Arabic proverbs classification, and violent/nonviolent video categorization using YouTube video content and Arabic comments. He has also published a number of research papers in various conferences and journals.

Eihab Bashier Mohammed Bashier earned a BSc and an MSc at the University of Khartoum, Sudan. He obtained a postgraduate diploma in mathematical sciences from the African Institute of Mathematical Sciences, Stellenbosch University, South Africa. He then earned a PhD at the University of the Western Cape in South Africa. He is an associate professor of applied mathematics at the University of Khartoum, Sudan. Recently, he has joined the Department of Mathematics, Physics, and Statistics of Qatar University. His research interests include numerical methods for differential equations with applications to biology, and information and computer security. Dr. Bashier supervises postgraduate students. He has also published several research articles in international journals. Dr. Bashier received the African Union and the Third World Academy of Science (AU-TWAS) Young Scientists' National Award in Basic Sciences, Technology, and Innovation in 2011. He is a reviewer for many international journals and is an IEEE member.

Introduction

Since their evolution, humans have been using many types of tools to accomplish various tasks. The creativity of the human brain led to the invention of different machines. These machines made the human life easy by enabling people to meet various life needs, including travelling, industries, constructions, and computing.

Despite rapid developments in the machine industry, intelligence has remained the fundamental difference between humans and machines in performing their tasks. A human uses his or her senses to gather information from the surrounding atmosphere; the human brain works to analyze that information and takes suitable decisions accordingly. Machines, in contrast, are not intelligent by nature. A machine does not have the ability to analyze data and take decisions. For example, a machine is not expected to understand the story of Harry Potter, jump over a hole in the street, or interact with other machines through a common language.

The era of intelligent machines started in the mid-twentieth century when Alan Turing thought whether it is possible for machines to think. Since then, the artificial intelligence (AI) branch of computer science has developed rapidly. Humans have had the dreams to create machines that have the same level of intelligence as humans. Many science fiction movies have expressed these dreams, such as *Artificial Intelligence*; *The Matrix*; *The Terminator*; *I, Robot*; and *Star Wars*.

The history of AI started in the year 1943 when Waren McCulloch and Walter Pitts introduced the first neural network model. Alan Turing introduced the next noticeable work in the development of the AI in 1950 when he asked his famous question: can machines think? He introduced the B-type neural networks and also the concept of test of intelligence. In 1955, Oliver Selfridge proposed the use of computers for pattern recognition.

In 1956, John McCarthy, Marvin Minsky, Nathan Rochester of IBM, and Claude Shannon organized the first summer AI conference at Dartmouth College, the United States. In the second Dartmouth conference, the term *artificial intelligence* was used for the first time. The term *cognitive science* originated in 1956, during a symposium in information science at the MIT, the United States.

Rosenblatt invented the first perceptron in 1957. Then in 1959, John McCarthy invented the LISP programming language. David Hubel and Torsten Wiesel proposed the use of neural networks for the computer vision in 1962. Joseph Weizenbaum developed the first expert system *Eliza* that could diagnose a disease from its symptoms. The National Research Council (NRC) of the United States founded the Automatic Language Processing Advisory Committee (ALPAC) in 1964 to advance the research in the natural language processing. But after many years, the two organizations terminated the research because of the high expenses and low progress.

Marvin Minsky and Seymour Papert published their book *Perceptrons* in 1969, in which they demonstrated the limitations of neural networks. As a result, organizations stopped funding research on neural networks. The period from 1969 to 1979 witnessed a growth in the research of knowledge-based systems. The developed programs Dendral and Mycin are examples of this research. In 1979, Paul Werbos proposed the first efficient neural network model with backpropagation. However, in 1986, David Rumelhart, Geoffrey Hinton, and

Ronald Williams discovered a method that allowed a network to learn to discriminate between nonlinear separable classes, and they named it *backpropagation*.

In 1987, Terrence Sejnowski and Charles Rosenberg developed an artificial neural network NETTalk for speech recognition. In 1987, John H. Holland and Arthur W. Burks invented an adapted computing system that is capable of learning. In fact, the development of the theory and application of genetic algorithms was inspired by the book *Adaptation in Neural and Artificial Systems*, written by Holland in 1975. In 1989, Dean Pomerleau proposed ALVINN (autonomous land vehicle in a neural network), which was a three-layer neural network designed for the task of the road following.

In the year 1997, the Deep Blue chess machine, designed by IBM, defeated Garry Kasparov, the world chess champion. In 2011, Watson, a computer developed by IBM, defeated Brad Rutter and Ken Jennings, the champions of the television game show *Jeopardy!*

The period from 1997 to the present witnessed rapid developments in reinforcement learning, natural language processing, emotional understanding, computer vision, and computer hearing.

The current research in machine learning focuses on computer vision, hearing, natural languages processing, image processing and pattern recognition, cognitive computing, knowledge representation, and so on. These research trends aim to provide machines with the abilities of gathering data through senses similar to the human senses and then processing the gathered data by using the computational intelligence tools and machine learning methods to conduct predictions and making decisions at the same level as humans.

The term *machine learning* means to enable machines to learn without programming them explicitly. There are four general machine learning methods: (1) supervised, (2) unsupervised, (3) semi-supervised, and (4) reinforcement learning methods. The objectives of machine learning are to enable machines to make predictions, perform clustering, extract association rules, or make decisions from a given dataset.

This book focuses on the supervised and unsupervised machine learning techniques. We provide a set of MATLAB programs to implement the various algorithms that are discussed in the chapters.

Chapter 1

Introduction to Machine Learning

1.1 Introduction

Learning is a very personalized phenomenon for us. Will
Durant in his famous book, *The Pleasures of Philosophy*, won-
dered in the chapter titled "Is Man a Machine?" when he wrote
such classical lines:

> Here is a child; … See it raising itself for the first
> time, fearfully and bravely, to a vertical dignity; why
> should it long so to stand and walk? Why should it
> tremble with perpetual curiosity, with perilous and
> insatiable ambition, touching and tasting, watch-
> ing and listening, manipulating and experimenting,
> observing and pondering, *growing*—till it weighs the
> earth and charts and measures the stars?… [1]

Nevertheless, learning is not limited to humans only. Even
the simplest of species such as amoeba and paramecium
exhibit this phenomenon. Plants also show intelligent

behavior. Only nonliving things are the natural stuffs that are not involved in learning. Hence, it seems that *living* and *learning* go together. In nature-made nonliving things, there is hardly anything to learn. Can we introduce learning in human-made nonliving things that are called *machines*? Enabling a machine capable of learning like humans is a dream, the fulfillment of which can lead us to having *deterministic machines* with *freedom* (or illusion of freedom in a sense). During that time, we will be able to happily boast that our humanoids resemble the image and likeliness of *humans* in the guise of machines.

1.2 Preliminaries

Machines are by nature not intelligent. Initially, machines were designed to perform specific tasks, such as running on the railway, controlling the traffic flow, digging deep holes, traveling into the space, and shooting at moving objects. Machines do their tasks much faster with a higher level of precision compared to humans. They have made our lives easy and smooth.

The fundamental difference between humans and machines in performing their work is intelligence. The human brain receives data gathered by the five senses: vision, hearing, smell, taste, and tactility. These gathered data are sent to the human brain via the neural system for perception and taking action. In the perception process, the data is organized, recognized by comparing it to previous experiences that were stored in the memory, and interpreted. Accordingly, the brain takes the decision and directs the body parts to react against that action. At the end of the experience, it might be stored in the memory for future benefits.

A machine cannot deal with the gathered data in an intelligent way. It does not have the ability to analyze data for classification, benefit from previous experiences, and store the new experiences to the memory units; that is, machines do not learn from experience.

Although machines are expected to do mechanical jobs much faster than humans, it is not expected from a machine to: understand the play *Romeo and Juliet*, jump over a hole in the street, form friendships, interact with other machines through a common language, recognize dangers and the ways to avoid them, decide about a disease from its symptoms and laboratory tests, recognize the face of the criminal, and so on. The challenge is to make *dumb* machines learn to cope correctly with such situations. Because machines have been originally created to help humans in their daily lives, it is necessary for the machines to *think, understand* to solve problems, and *take* suitable decisions akin to humans. In other words, we need *smart* machines. In fact, the term *smart machine* is symbolic to machine learning success stories and its future targets. We will discuss the issue of smart machines in Section 1.4.

The question of whether a machine can think was first asked by the British mathematician Alan Turing in 1955, which was the start of the artificial intelligence history. He was the one who proposed a *test* to measure the performance of a machine in terms of intelligence. Section 1.4 also discusses the progress that has been achieved in determining whether our machines can pass the Turing test.

Computers are machines that follow programming instructions to accomplish the required tasks and help us in solving problems. Our brain is similar to a CPU that solves problems for us. Suppose that we want to find the smallest number in a list of unordered numbers. We can perform this job easily. Different persons can have different methods to do the same job. In other words, different persons can use different *algorithms* to perform the same task. These methods or algorithms are basically a sequence of instructions that are executed to reach from one state to another in order to produce output from input.

If there are different algorithms that can perform the same task, then one is right in questioning which algorithm is better. For example, if two programs are made based on two different algorithms to find the smallest number in an unordered list, then for the same list of unordered number (or same set of input) and on the same machine, one measure of efficiency can be speed or quickness of program and another can be minimum memory usage. Thus, time and space are the usual measures to test the efficiency of an algorithm. In some situations, time and space can be inter-related, that is, the reduction in memory usage leading to fast execution of the algorithm. For example, an efficient algorithm enabling a program to handle full input data in cache memory will also consequently allow faster execution of program.

1.2.1 Machine Learning: Where Several Disciplines Meet

Machine learning is a branch of *artificial intelligence* that aims at enabling machines to perform their jobs skillfully by using intelligent software. The statistical learning methods constitute the backbone of intelligent software that is used to develop machine intelligence. Because machine learning algorithms require data to learn, the discipline must have connection with the discipline of database. Similarly, there are familiar terms such as Knowledge Discovery from Data (KDD), data mining, and pattern recognition. One wonders how to view the big picture in which such connection is illustrated.

SAS Institute Inc., North Carolina, is a developer of the famous analytical software Statistical Analysis System (SAS). In order to show the connection of the discipline of machine learning with different related disciplines, we will use the illustration from SAS. This illustration was actually used in a data mining course that was offered by SAS in 1998 (see Figure 1.1).

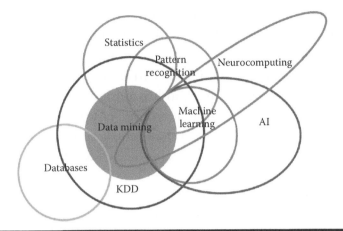

Figure 1.1 **Different disciplines of knowledge and the discipline of machine learning. (From Guthrie, Looking backwards, looking forwards: SAS, data mining and machine learning, 2014, http://blogs.sas.com/ content/subconsciousmusings/2014/08/22/looking-backwards-looking-forwards-sas-data-mining-and-machine-learning/2014. With permission.)**

In a 2006 article entitled "The Discipline of Machine Learning," Professor Tom Mitchell [3, p.1] defined the discipline of machine learning in these words:

> Machine Learning is a natural outgrowth of the **intersection of Computer Science and Statistics**. We might say the defining question of Computer Science is 'How can we build machines that solve problems, and which problems are inherently tractable/intractable?' The question that largely defines Statistics is 'What can be inferred from data plus a set of modeling assumptions, with what reliability?' The defining question for Machine Learning builds on **both**, but it is a distinct question. Whereas Computer Science has focused primarily on how to manually program computers, Machine Learning focuses on the question of **how to get computers to program themselves (from experience plus some initial structure)**. Whereas Statistics

has focused primarily on what conclusions can be inferred from data, Machine Learning incorporates additional questions about what computational architectures and algorithms can be used to most effectively capture, store, index, retrieve and merge these data, how multiple learning subtasks can be orchestrated in a larger system, and questions of computational tractability [emphasis added].

There are some tasks that humans perform effortlessly or with some efforts, but we are unable to explain how we perform them. For example, we can recognize the speech of our friends without much difficulty. If we are asked how we recognize the voices, the answer is very difficult for us to explain. Because of the lack of understanding of such phenomenon (speech recognition in this case), we cannot craft algorithms for such scenarios. *Machine learning* algorithms are helpful in bridging this gap of understanding.

The idea is very simple. We are not targeting to understand the underlying processes that help us learn. We write computer programs that will make machines learn and enable them to perform tasks, such as prediction. The goal of learning is to construct a *model* that takes the input and produces the desired result. Sometimes, we can understand the model, whereas, at other times, it can also be like a black box for us, the working of which cannot be intuitively explained. The model can be considered as an *approximation* of the process we want machines to mimic. In such a situation, it is possible that we obtain errors for some input, but most of the time, the model provides correct answers. Hence, another measure of performance (besides performance of metrics of speed and memory usage) of a machine learning algorithm will be the *accuracy* of results. It seems appropriate here to quote another statement about learning of computer program from Professor Tom Mitchell from Carnegie Mellon University [4, p.2]:

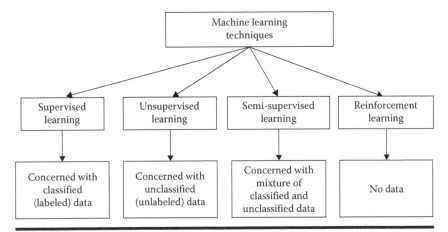

Figure 1.2 Different machine learning techniques and their required data.

> A computer program is said to learn from experi-
> ence **E** with respect to some class of tasks **T** and
> performance measure **P**, if its performance at tasks
> in **T**, as measured by **P**, improves with experience **E**.

The subject will be further clarified when the issue will be discussed with examples at their relevant places. However, before the discussion, a few widely used terminologies in the machine learning or data mining community will be discussed as a prerequisite to appreciate the examples of machine learning applications. Figure 1.2 depicts four machine learning techniques and describes briefly the nature of data they require. The four techniques are discussed in Sections 1.2.2 through 1.2.5.

1.2.2 Supervised Learning

In supervised learning, the target is to infer a function or mapping from training data that is *labeled*. The training data consist of input vector **X** and output vector **Y** of labels or tags. A label or tag from vector **Y** is the *explanation* of its respective input example from input vector **X**. Together they form a *training example*. In other words, training data comprises training examples. If the labeling does not exist for input vector **X**, then **X** is *unlabeled data.*

Why such learning is called *supervised learning?* The output vector **Y** consists of labels for each training example present in the training data. These labels for output vector are provided by the supervisor. Often, these supervisors are humans, but machines can also be used for such labeling. Human judgments are more expensive than machines, but the higher error rates in data labeled by machines suggest superiority of human judgment. The manually labeled data is a precious and reliable resource for supervised learning. However, in some cases, machines can be used for reliable labeling.

Example

Table 1.1 demonstrates five unlabeled data examples that can be labeled based on different criteria.

The second column of the table titled, "Example judgment for labeling" expresses possible criterion for each data example. The third column describes possible labels after the application of judgment. The fourth column informs which actor can take the role of the supervisor.

In all first four cases described in Table 1.1, machines can be used, but their low accuracy rates make their usage questionable. Sentiment analysis, image recognition, and speech detection technologies have made progress in past three decades but there is still a lot of room for improvement before we can equate them with humans' performance. In the fifth case of tumor detection, even normal humans cannot label the X-ray data, and expensive experts' services are required for such labeling.

Two groups or categories of algorithms come under the umbrella of supervised learning. They are

1. Regression
2. Classification

Table 1.1 Unlabeled Data Examples along with Labeling Issues

Unlabeled Data Example	Example Judgment for Labeling	Possible Labels	Possible Supervisor
Tweet	Sentiment of the tweet	Positive/ negative	Human/ machine
Photo	Contains house and car	Yes/No	Human/ machine
Audio recording	The word football is uttered	Yes/No	Human/ machine
Video	Are weapons used in the video?	Violent/ nonviolent	Human/ machine
X-ray	Tumor presence in X-ray	Present/ absent	Experts/ machine

1.2.3 Unsupervised Learning

In unsupervised learning, we lack *supervisors* or training data. In other words, all what we have is unlabeled data. The idea is to find a hidden structure in this data. There can be a number of reasons for the data not having a label. It can be due to unavailability of funds to pay for manual labeling or the inherent nature of the data itself. With numerous data collection devices, now data is collected at an unprecedented rate. The variety, velocity, and the volume are the dimensions in which *Big Data* is seen and judged. To get something from this data without the supervisor is important. This is the challenge for today's machine learning practitioner.

The situation faced by a machine learning practitioner is somehow similar to the scene described in *Alice's Adventures in Wonderland* [5, p.100], an 1865 novel, when Alice looking to go *somewhere*, talks to the Cheshire cat.

… She went on. "Would you tell me, please, **which way** I ought to go from here?"

"That depends a good deal on **where** you want to get to," said the Cat.

"I don't much care **where—**" said Alice.

"Then it **doesn't matter** which way you go," said the Cat.

"—so long as I get ***somewhere***," Alice added as an explanation.

"Oh, you're sure to do that," said the Cat, "if you only **walk long enough**."

In the machine learning community, probably *clustering* (an unsupervised learning algorithm) is analogous to the *walk long enough* instruction of the Cheshire cat. The *somewhere* of Alice is equivalent to *finding regularities in the input*.

1.2.4 Semi-Supervised Learning

In this type of learning, the given data are a mixture of classified and unclassified data. This combination of labeled and unlabeled data is used to generate an appropriate model for the classification of data. In most of the situations, labeled data is scarce and unlabeled data is in abundance (as discussed previously in unsupervised learning description). The target of semi-supervised classification is to learn a model that will predict classes of future test data better than that from the model generated by using the labeled data alone. The way we learn is similar to the process of semi-supervised learning. A child is supplied with

1. Unlabeled data provided by the environment. The surroundings of a child are full of unlabeled data in the beginning.

2. Labeled data from the supervisor. For example, a father teaches his children about the names (labels) of objects by pointing toward them and uttering their names.

Semi-supervised learning will not be discussed further in the book.

1.2.5 Reinforcement Learning

The reinforcement learning method aims at using observations gathered from the interaction with the environment to take actions that would maximize the reward or minimize the risk.

In order to produce intelligent programs (also called *agents*), reinforcement learning goes through the following steps:

1. Input state is observed by the agent.
2. Decision making function is used to make the agent perform an action.
3. After the action is performed, the agent receives reward or reinforcement from the environment.
4. The state-action pair information about the reward is stored.

Using the stored information, policy for particular state in terms of action can be fine-tuned, thus helping in optimal decision making for our agent.

Reinforcement learning will not be discussed further in this book.

1.2.6 Validation and Evaluation

Assessing whether the model learnt from machine learning algorithm is good or not, needs both validation and evaluation. However, before discussing these two important terminologies, it is interesting to mention the writings of Plato (the great philosopher) regarding this issue. The excerpt from his approach is given in Box 1.1 to introduce readers to this interesting debate.

BOX 1.1 PLATO ON STABILITY OF BELIEF

Plato's ethics is written by Terence Irwin, professor of the history of philosophy at the University of Oxford. In section 101 titled "Knowledge, Belief and Stability," there is an interesting debate about *wandering* of beliefs. The following are excerpts from the book.

> Plato also needs to consider the different circumstance that might cause true beliefs to wander away ... Different demands for stability might rest on different standards of reliability. If, for instance, I believe that these animals are sheep, and they are sheep, then my belief is reliable for these animals, and it does not matter if I do not know what makes them sheep. If, however, I cannot tell the difference between sheep and goat and do not know why these animals are sheep rather than goats, my ignorance would make a difference if I were confronted with goats. If we are concerned about 'empirical reliability' (the tendency to be right in empirically likely conditions), my belief that animals with a certain appearance are sheep may be perfectly reliable (if I can be expected not to meet any goats). If we are concerned about 'counterfactual reliability' (the tendency to be right in counterfactual, and not necessarily empirically likely, conditions), my inability to distinguish sheep from goats make

(Continued)

BOX 1.1 (CONTINUED)   PLATO ON STABILITY OF BELIEF

my belief unreliable that animals with certain appearance are sheep. In saying that my belief about sheep is counterfactually unreliable, we point out that my reason for believing that these things are sheep is mistaken, even though the mistake makes no difference to my judgements in actual circumstances.

When Plato speaks of a given belief 'wandering', he describes a fault that we might more easily recognize if it were described differently. If I identify sheep by features that do not distinguish them from goats, then I rely on false principles to reach the true belief 'this is a sheep' in an environment without goats. If I rely on the same principles to identify sheep in an environment that includes goats, I will often reach the false belief 'this is a sheep 'when I meet a goat. We may want to describe these facts by speaking of three things: (1) the true token belief 'this is a sheep' (applied to a sheep in the first environment), (2) the false token belief 'this is a sheep' (applied to a goat in the second environment), and (3) the false general principle that I use to identify sheep in both environments.

If one claims that for a particular training data the function fits perfectly, then for the machine learning community, this claim is not enough. They will immediately ask about the performance of function on testing data.

A function fitting perfectly on training data needs to be examined. Sometimes, it is the phenomenon of *overfitting* that will give best performance on training data, and when yet-unseen labeled data will be used to test them, they will fail miserably. To avoid overfitting, it is common practice to divide the labeled data into two parts:

1. Training data
2. Testing data

A training set is used to build the model and testing set is used to validate the built model. In hold out testing/validation, one is expected to hold out part of the data for testing. Larger portion of the data is used for model training purpose, and the test metrics of the model are tested on holdout data.

The technique of cross-validation is useful when the available training dataset is quite small and one cannot afford to hold out part of the data just for validation purposes. In k-fold cross-validation, the available dataset is divided into k equal folds. Each of these k folds are treated as holdout datasets, and the training of the model is performed on rest of the $k - 1$ folds. The performance of the model is judged on the basis of holdout fold. The average of performance on all k folds is the overall performance of model.

1.3 Applications of Machine Learning Algorithms

Machine learning has proven itself to be the answer to many real-world challenges, but there are still a number of problems for which machine learning breakthrough is required. The need was felt by the cofounder and ex-chairman of Microsoft, Bill Gates, and was translated into the following wordings on one occasion [6]:

A breakthrough in machine learning would be worth ten Microsofts

In this section, we will discuss some applications of machine learning with some examples.

1.3.1 Automatic Recognition of Handwritten Postal Codes

Today, in order to communicate, we use a variety of digital devices. However, the postal services still exist, helping us send our mails, gifts, and important documents to the required destination. The way machine learning has benefited this sector can be understood by citing the example of the US Postal Service.

The US Postal Service was able to exploit the potentials of machine learning in the 1960s when they successfully used machines to automatically read the city/state/ZIP code line of typed addresses to sort letters. Optical character recognition (OCR) technology was able to correctly interpret the postal address using machine learning algorithm. According to the author:

> The images that consist of typed, handwritten or printed content of text are readable for humans. In order to make such text content readable for machines, Optical character recognition technology is used.
>
> A scanned text document in an image format like bitmap is nothing but a picture of the text. OCR software analyzes the image and attempts to identify alphabetic letter and numeric digit. When it is successful in recognizing a character, it converts into machine-encoded text. This machine-encoded text can be electronically edited, searched,compressed and can be used as input for applications like automatic translation, text-to-speech and text mining. In the presence of accurate OCR, data entry becomes simple, faster and cheaper.

Figure 1.3   Image of OCR camera used by the US Postal Service.

In 1900, the post office was handling 7.1 billion pieces of mail per year.* All these were being done without cars or sophisticated machinery.

In 2006,† the US Postal Service sorted and delivered more than 213 billion pieces of mail, about 40% of the world's total mail volume and more than any other postal administration in the world. This enormous herculean service is provided by the US Postal Service with the help of machines.

OCR has been successful in bringing a new revolution in the efficiency of postal system. The OCR camera similar to the one shown in Figure 1.3 helped in forming the connection between the physical mail and the information system that directs it to its destination. Now improved OCR technology accompanied with other mail processing services is able to enhance the efficiency of different countries' postal services. According to the US Postal Service

---

* http://www.theatlantic.com/technology/archive/2011/12/tech-has-saved-the-postal-service-for-200-years-today-it-wont/249946/#slide10.
† https://about.usps.com/publications/pub100/pub100_055.htm.

website [7], "the Postal Service is the world leader in optical character recognition technology with machines reading nearly 98% of all hand-addressed letter mail and 99.5% of machine-printed mail."

Google is now providing free service to convert the text image to text documents for 200 languages in more than 25 writing systems. Google accomplished this by using hidden Markov models and treated the input as a whole sequence, rather than first trying to break it apart into pieces. The list of supported languages can be found at the Google website.* One can imagine the complexities involved in such a task. A simple problem is language identification. There no longer exists a hidden supposition that the language of the document to be processed is already known. If the language is identified wrongly, it means that we should expect a poor performance from the OCR technology.

The OCR technology is one of the applications of pattern recognition, a branch of machine learning. The focus of pattern recognition is to recognize pattern and regularities in data. The data can be text, speech, and/or image. The OCR example is the one in which input data is in the form of an image. Another example of the application of pattern recognition using image data is computer-aided diagnosis. We will discuss some of its applications in Section 1.3.2.

1.3.2 Computer-Aided Diagnosis

Pattern recognition algorithms used in computer-aided diagnosis can assist doctors in interpreting medical images in a relatively short period. Medical images from different medical tests such as X-rays, MRI, and ultrasound are the sources of data describing a patient's condition.

---

* https://support.google.com/drive/answer/176692?hl=en.

The responsibility of a radiologist is to analyze and evaluate the output of these medical tests that are in the form of a digital image. The short time constraint requires that the radiologist be assisted by machine. Computer-aided diagnosis uses pattern recognition techniques from machine learning to identify suspicious structures in the image. How does an algorithm catch suspicious structure? Supervised learning is done to perform this task. Few thousand labeled images are given to the machine learning algorithm, such as Bayesian classifier, artificial neural network, radial basis function network, and support vector machine. The resulting classifier is expected to classify new medical images correctly.

Mistakes in diagnosis by the machine learning algorithm can bring disaster for a family. The fault can cause damage to a person in monetary terms and it can risk his/her life, too.

The following are two such examples:

1. Suppose our classifier detects breast cancer in a patient who actually had no such disease. The results obtained by the classifier will create harmful psychological conditions for the patient. In order to confirm the result of the classifier, further tests can result in monetary losses for the patient.

2. Suppose our classifier does not detect breast cancer in patient who actually has such a disease. This will lead to wrong medical treatment and can threaten the life of the patient in near or far future. In order to avoid such mistakes, the complete substitution of doctor with technology is not recommended. The role of technology should be supportive. It should be the doctor(generally a radiologist) who must take the responsibility of the final interpretation of medical image.

Computer-aided diagnosis is assisting medical doctors/ radiologists in the diagnosis of a number of health problems. Few examples are as follows:

- Pathological brain detection
- Breast cancer
- Lung cancer
- Colon cancer
- Prostate cancer
- Bone metastases
- Coronary artery disease
- Congenital heart defect
- Alzheimer's disease

1.3.3 Computer Vision

We want our robots to see and act appropriately after under-standing the situation. The cameras installed in a robot can provide images, but they will not help the robot recognize or interpret the image. Using pattern recognition, what type of learning can a robot perform? We begin with the discussion of the example of the event called *RoboCup*.

> *RoboCup*: "Robot Soccer World Cup" or RoboCup is an international robotic tournament of soccer. The officially declared goal of the project is very challenging and is stated as follows:

"By the middle of the 21st century, a team of fully autonomous humanoid robot soccer players shall win a soccer game, com-plying with the official rules of FIFA, against the winner of the most recent World Cup."* The RoboCup 2015 (China) attracted 175 intelligent sporting robot teams from 47 countries. In the largest adult size category of the event, the US team designed

---

* http://www.robocup.org/about-robocup/objective/.

Figure 1.4   US and Iranian robot teams competing for RoboCup 2014 final. (Courtesy of Reuters.)

by the University of Pennsylvania defeated the Iranian team with tough 5–4 goal results (Figure 1.4).

The autonomous robots are expected to cooperate with their other team members (that are also robots) in adversarial dynamic environment in order to win the match. They need to categorize objects and recognize activities. To perform these tasks, they get input from their cameras. These tasks lie purely in the pattern recognition domain, a branch of machine learning.

1.3.3.1 Driverless Cars

Autonomous cars with no drivers are also one of the applications where *car vision* is actually made possible by advancement in the computer vision technology. In the industry, it is clear that there is ongoing competition to manufacture driverless cars running on the roads as early as possible. According to the BBC[*] report titled *Toyota promises driverless cars on roads by 2020*, different competitors

---

[*] http://www.bbc.com/news/technology-34464450.

are on the bandwagon and announcing their targets for driverless cars. The article states:

> **Toyota** is the latest car company to push forward with plans for an autonomous vehicle, offering fresh competition to Silicon Valley companies such as **Google**, **Cruise** and **Tesla**.
>
> Last week, **General Motors** said it was offering driverless rides to workers at its research and development facility in Warren, Michigan.
>
> **Nissan** has promised to put an automated car on Japan's roads as early as 2016.
>
> However, **Google** is already testing its self-driving cars on US city streets. And **Tesla** chief executive Elon Musk said in July his company was "almost ready" to make its cars go driverless on main roads and parallel-park themselves.

How these cars will accomplish this task? BBC article states the narrative of Toyota in following words:

> According to Toyota, the car "uses multiple external sensors to **recognise** nearby vehicles and hazards, and selects appropriate routes and lanes depending on the destination."

Based on these data inputs, it "'automatically operates the steering wheel, accelerator and brakes' to drive in much the same way as a person would drive." (Figure 1.5) [8].

The applications that are and will be using computer-vision-related technologies are very sensitive in nature. A driverless car accident can result in a tragedy for family or families. Similarly, another very sensitive area is the usage of computer vision technology in drones. The drones that are used in warfare can kill innocent people if algorithms behind the vision misbehave.

Figure 1.5   Toyota tested its self-driving Highway Teammate car on a public road. (Courtesy of BBC.)

1.3.3.2 Face Recognition and Security

Images from smart phones and CCTV cameras are now produced at an unprecedented rate. A problem pertinent to face recognition is to associate the face image to its respective identity. Building a classifier for this task is not a trivial job, because there are too many classes involved with multiple image-related problems. Face recognition can help security agencies to use a large amount of data from different sources to automatically find what is very difficult for humans to do manually.

1.3.4 Speech Recognition

The field of speech recognition aims to develop methodologies and technologies that enable computers to recognize and translate spoken language into text. Stenography (writing in shorthand) is no longer required. Automatic transcription of speech into text has found its way in areas such as video captioning and court reporting. This technology can help people with disabilities. With the passage of time, the accuracy of speech recognition engines is increasing. There is no doubt that voice-controlled programs such as Apple's Siri, Google

Now, Amazon's Alexa, and Microsoft's Cortana do not always understand our speech, but things are likely to be improved in the near future.

1.3.5 Text Mining

The examples that we have studied up until now are basically using image or voice data for learning. We have another source of learning, that is, text data. It was observed that most of the enterprise-related information is stored in text format. The challenge was how to use this unstructured data or text. The earliest definition or function of business intelligence system given by H.P. Luhn [9] in the IBM journal is as follows:

> … utilize data-processing machines for **auto-abstracting** and **auto-encoding** of documents and for creating interest profiles for each of the 'action points' in an organization. Both incoming and internally generated documents are automatically abstracted, characterized by a word pattern, and sent automatically to appropriate action points.

Another venue where the unstructured data or text is available in abundance for researchers is social media. Social media is the place where we can see the production of text data at an unprecedented level. The sharing of personal experiences in the form of text has provided stakeholders, such as business, the opportunity to analyze and use them for beneficial purpose.

Text mining is helpful in a number of applications including

■ Business intelligence
■ National security
■ Life sciences
■ Those related to sentiment classification
■ Automated placement of advertisement

- Automated classification of news articles
- Social media monitoring
- Spam filter

1.3.5.1 Where Text and Image Data Can Be Used Together

It is possible that in order to solve a particular problem, both text and image data are used. For example, the problem of author identification for a particular written corpus of data can be solved in two ways:

1. *Handwriting detection*: The known corpus of hand-written data can be used to make a classifier that can assign a document to an author based on different features.
2. *Writing style detection*: This is a text mining problem. We want to find features that are related to a peculiar author using known documents attributed to the author. These features can be used to build a classifier that can identify whether the particular document belongs to the author or not.

It is possible that the two classifiers are joined together to develop a new classifier with improved performance for author identification.

Another area where such data can be helpful in solving the problem is in the identification of unwanted material in a video. In order to identify unwanted material, we can approach the problem in two ways:

1. Use video images and apply machine learning techniques on image data to make a model to identify unwanted material in the video.

2. Use comments from social media related to video to understand the content of the video by making a model that can predict the presence or absence of unwanted material in the video.

Once again, the two classifiers can be combined to improve the performance of the system.

1.4 The Present and the Future

1.4.1 Thinking Machines

The question of whether a machine can think was first asked by the British mathematician Alan Turing in 1955, which was the start of the artificial intelligence history. He was the one who proposed a *test* to measure the machine's performance in terms of intelligence.

In 2014, a chatbot was able to pass this Turing test (see Box 1.2 for further details). A chatbot is a computer program that simulates an intelligent conversation with one or more human users. This conversation can be performed via audio or text communication methods. Box 1.3 describes another interesting event in which one of the judges of the annual Loebner Prize* 2015 discusses the deficiencies of chatbots. We have included a full transcript of the chat between one of the judges and the 2015 winner chatbot of Loebner Prize in the Appendix I. The transcript will help readers understand how chatbots try to dodge the judges when faced with difficult questions.

Researchers at Google have programmed an advanced type of chatbot that is able to learn from training data comprising of examples from dialogues. The two sources of training data were IT helpdesk troubleshooting dataset and movie transcript dataset.

---

* https://en.wikipedia.org/wiki/Loebner_Prize (accessed on September 20, 2015).

BOX 1.2   TURING TEST PASSED BY CHATBOT NAMED EUGENE

The Turing test is based on twentieth century-mathematician and code-breaker Alan Turing's 1950 famous question and answer game, *Can Machines Think?* The experiment investigates whether people can detect if they are talking to machines or humans. If a computer is mistaken for a human more than 30% of the time during a series of a five-minute keyboard conversation, the machine passes the test.

In 2014, a computer program Eugene Goostman passed the Turing test for the first time during *Turing Test 2014* held at the renowned Royal Society in London on June 7, 2014. Eugene managed to convince 33% of the human judges (30 judges took part) that it was human.

Source: http://www.reading.ac.uk/news-and-events/releases/PR583836.aspx

BOX 1.3   THE DIFFERENCE BETWEEN CONVERSATION WITH HUMAN AND A MACHINE

In the Tech section of the BBC website, the story appeared with the title of *AI bots try to fool human judges,* describing the live reporting of annual Loebner Prize 2015. One of the judges of the event, who had to evaluate the intelligence of a chatbot, was BBC technology correspondent Rory Cellan-Jones. The full transcript of his conversation with the 2015 prize winner, the Chatbot Rose, is given on the BBC website. The comments from Rory after the whole experience are as follows:

(Continued)

BOX 1.3 (CONTINUED) THE DIFFERENCE BETWEEN CONVERSATION WITH HUMAN AND A MACHINE

I feel as though I'm a psychiatrist who has just spent two hours delving into the innermost thoughts of four pairs of patients.

Being a judge in the Loebner Prize has made me think about how conversations work—and what it means to be a human conversationalist.

I quickly latched on to a simple technique for spotting the bot—be a messy human chatter.

The bots could cope with simple questions—where do you live, what do you do, how did you get here.

But the minute I started musing on London house prices, how to deal with slugs in your garden, they just fell apart.

Their technique was to try to take the conversation in another direction, ignoring what I was saying.

So, it took me no more than two or three questions to work out which was the bot and which the human.

My conclusion—it will take some time before a computer passes the Turing Test. The humans are just much more interesting to talk to.

Source: http://www.bbc.com/news/ live/technology-34281198

They trained their chatbot with language model based on recurrent neural network. It means that these are not just canned answers that are given by chatbots seeing some patterns in human chats. Some of the interesting and artistic answers by the chatbot from Google are available in the research paper[*] titled, "A neural conversational model" [10]. The researchers admitted the limitation of the work in their research paper that the chatbot was unable to have a realistic conversation currently and, hence passed the Turing test; however proper answers to many different types of questions without rules is a surprising discovery. We have included different conversations of this learning chatbot in the Appendix II.

1.4.2 Smart Machines

The dream of machines appearing as smart as humans is still far from being realized. In general, a smart machine is an *intelligent* system that uses equipment such as sensors, RFID, a Wi-Fi, or cellular communications link to receive data and interpret it to make decisions. They use machine learning algorithms to accomplish tasks usually performed by humans in an order to enhance efficiency and productivity.

Gartner, Inc.[†], Stanford, California, is an American information technology (IT) research and advisory firm providing technology-related insight targeting CIOs and senior IT leaders by disseminating their research in a number of ways such as Gartner symposiums. Gartner symposium/ITxpo attracts thousands of CIOs from the industry. Gartner's analyst Kenneth F. Brant has given two criteria for a true smart machine.

A true smart machine meets two criteria[‡]:

1. First, a smart machine does something that no machine was ever thought to be able to do. Using that yardstick, a drone delivering a package—a model being

---

[*] http://arxiv.org/pdf/1506.05869v2.pdf.
[†] http://www.gartner.com/technology/about.jsp.
[‡] http://www.forbes.com/sites/emc/2014/01/09/smart-machines-shaping-the-workforce-of-the-future/.

contemplated by Amazon—would qualify as a smart machine
2. Machine is capable of learning. Using the second criterion for a true smart—the delivery drone fails the test

Yet that same delivery drone–regardless of how smart it is—could still have a significant effect on productivity and employment in the shipping industry.

Smart machines were one of the top 10 technologies and trends that were predicted to be strategic for most organizations in 2014 as well as 2015 by Gartner, Inc. The prediction for 2014 placed *smart machines* in the category of *future disruption* along with the technology of 3D printers. Smart machines were again present in the prediction for 2015 in the category of *intelligence everywhere* (Figure 1.6).

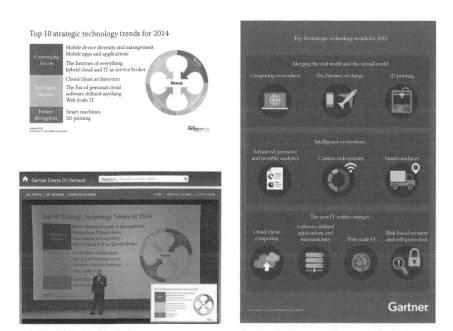

Figure 1.6   The top 10 strategic technologies in years 2014 and 2015.

In their prediction related to smart machines for 2014 and 2015, the following statements were made:

■ By 2015, there will be more than 40 vendors with commercially available managed services offerings leveraging smart machines and industrialized services. By 2018, the total cost of ownership for business operations will be *reduced by 30%* through smart machines and industrialized services.

■ Through 2020, the smart machine era will blossom with a proliferation of contextually aware, intelligent personal assistants, smart advisors (such as IBM Watson), advanced global industrial systems, and public availability of early examples of autonomous vehicles. The smart machine era will be the most disruptive in the history of IT.

In a recent report, *Cool Vendors in Smart Machines, 2014*, Gartner named three well-known examples of smart machines, including IBM's Watson, Google Now, and Apple's Siri. We will discuss few of the smart machines mentioned in the above predictions later in this chapter, but before that we will discuss *Deep Blue*, a chess-playing computer developed by IBM.

1.4.3 Deep Blue

In May 1997, IBM's Deep Blue became the first computer system to defeat the then-chess world champion Garry Kasparov in a match. The brute force of the computing power due to specialized hardware made Deep Blue capable of evaluating 200 million positions per second. The 259th most powerful supercomputer of 1997 was able to defeat the human world champion of chess. It was a historical achievement for the artificial intelligence community. How Deep Blue was able to evaluate the situation on the chess board? The answer to this question is as follows [11]:

> Deep Blue's evaluation function was initially written in a generalized form, with many to-be-determined parameters (e.g. how important is a safe king position compared to a space advantage in the center, etc.). The optimal values for these parameters were then determined by the system itself, by analyzing thousands of master games. The evaluation function had been split into 8,000 parts, many of them designed for special positions. In the opening book there were over 4,000 positions and 700,000 grandmaster games. The endgame database contained many six piece endgames and five or fewer piece positions

In 1997, Deep Blue was a dedicated supercomputer against humans. Now the focus of research in the chess domain is to improve software efficiency, so that less powerful hardware is enough for the task. In 2006, chess program named *Deep Fritz* played a chess match against world champion *Vladimir Kramnik*. The program was executed on a personal computer containing two Intel Core 2 Duo CPUs. The program was capable of evaluating only 8 million positions per second as compared to the 200 million positions per second evaluation power of Deep Blue.

1.4.4 IBM's Watson

It was named after the first CEO of IBM, Thomas J. Watson. IBM's Watson is a wonderful machine that is capable of answering the questions posed in natural language.

Whether you call it a supercomputer, a cognitive computing system, or simply a question answering matching system—IBM Watson is perhaps the most well-known example of artificial intelligence in use today.

Watson gained its worldwide fame by receiving the first prize inquiz show "Jeopardy!". With its supercomputing and AI power, Watson is able to help different industries by powering different types of practical applications. The industries benefiting from Watson include healthcare, finance, legal, and retail sector.

1.4.5 Google Now

Google's innovation "Google Now" is another landmark for machine learning world. It is a personal assistant with an element of smartness and intelligence in it. The functions of Google Now include answering questions, making recommendations, and performing actions by assigning requests to a set of web services. With it, users can use voice commands to create reminders and get help with trivia questions. The proactive program observes the search habits of the users and uses them to predict the information that may be useful for users and delivers it to them.

1.4.6 Apple's Siri

Siri (speech interpretation and recognition interface) is a widely used intelligent personal assistant by Apple Inc. Siri supports a number of languages including English, Spanish, French, German, Italian, Japanese, Korean, Mandarin, Russian, Turkish, and Arabic. Siri, just like any other personal assistant is updated to improve its response. Context understanding is very important. For example, if Siri is being told by a terrorist that he is going to blast a particular restuarant, Siri rather than showing the map of that restaurant, should respond by reporting such intention to some terrorism prevention center.

1.4.7 Microsoft's Cortana

Microsoft's Cortana is another intelligent personal assistant competing Google Now and Apple's Siri. Users will be soon able to use Skype to book trips, shop, and plan their schedules, by chatting with Cortana.

1.5 Objective of This Book

The objectives of this book are as follows:

- Explanation of the concepts of machine learning algorithms
- Demonstration of simple practical example(s) to make the reader understand each algorithm

We believe that this book will be a very useful resource for beginners as well as researchers and IT security professional.
We have divided our books into two sections.

1. Supervised Learning Algorithms
2. Unsupervised Learning Algorithms

In the first section, we will discuss following algorithms:

1. Decision trees
2. Rule-based algorithms
3. Naïve Bayesian algorithm
4. Nearest neighbor algorithm
5. Neural networks
6. Linear discriminant analysis
7. Support vector machine

In the second section, we will discuss following algorithms:

1. *K*-means
2. Gaussian mixture model
3. Hidden Markov model
4. Principal components analysis in the context of dimensionality reduction

References

1. Durant, W. Is man a machine? in *The Pleasures of Philosophy: A Survey of Human Life and Destiny.* New York: Simon and Schuster, 1953, p. 60.
2. Guthrie. Looking backwards, looking forwards: SAS, data mining and machine learning, 2014, http://blogs.sas.com/content/subconsciousmusings/2014/08/22/looking-backwards-looking-forwards-sas-data-mining-and-machine-learning/.
3. Mitchell, T. M. *The Discipline of Machine Learning*, Machine Learning Department technical report CMU-ML-06-108. Pittsburgh, PA: Carnegie Mellon University, July 2006.
4. Mitchell, T. M. *Machine Learning.* New York: McGraw-Hill, 1997.
5. Carroll, L. and Kelly, R. M. *Alice's Adventures in Wonderland.* Peterborough, ON: Broadview Press, 2000.
6. Bill Gates. AZQuotes.com, Wind and Fly LTD, 2016. http://www.azquotes.com/quote/850928 (accessed April 13, 2016).
7. United States Postal Service. https://about.usps.com/who-we-are/postal-facts/innovation-technologies.htm (accessed October 4, 2015).
8. Toyota Global Newsroom. http://http://newsroom.toyota.co.jp/en/detail/9753831 (accessed April 14, 2016).
9. Luhn, H. P. A business intelligence system, *IBM Journal* 2(4):314-319, 1958.
10. Vinyals, O. and Le, Q. V. A neural conversational model, *Proceedings of the 31st International Conference in Machine Learning*, Vol.37, arXiv:1506.05869v3, 2015.
11. Campbell, M., Hoane Jr. A. J., and Hsu, F.-H. Deep blue, *Artificial Intelligence* 134(1–2): 57-83, 2002.

SUPERVISED LEARNING ALGORITHMS

Introduction

In Section I, we will discuss the following algorithms:

1. Decision trees
2. Rule-based algorithms
3. Naïve Bayesian algorithm
4. Nearest neighbor algorithm
5. Neural networks
6. Linear discriminant analysis
7. Support vector machine

We will discuss the theoretical aspects of the above-mentioned algorithms as well as the practical MATLAB® implementations for simple examples. The idea is to provide the readers with simple examples, so that they can start their journey in learning these algorithms in an easy manner. We expect that this will help our readers to build strong foundations and give them insights to have better knowledge of the field.

Chapter 2

Decision Trees

2.1 Introduction

Decision tree (DT) is a statistical model that is used in classification. This machine-learning approach is used to classify data into classes and to represent the results in a flowchart, such as a tree structure [1]. This model classifies data in a dataset by flowing through a query structure from the root until it reaches the leaf, which represents one class. The root represents the attribute that plays a main role in classification, and the leaf represents the class. The DT model follows the steps outlined below in classifying data:

1. It puts all training examples to a root.
2. It divides training examples based on selected attributes.
3. It selects attributes by using some *statistical measures.*
4. Recursive partitioning continues until no training example remains, or until no attribute remains, or the remaining training examples belong to the same class.

In other words, we induce DT greedily in top-down fashion. This top-down induction of decision trees (TDIDTs) [2] is the most common strategy used to learn DTs from data.

DT learning is *supervised*, because it constructs DT from class-labeled training tuples. In this chapter, we will discuss two DT algorithms as follows:

1. ID3 (Iterative Dichotomiser 3)
2. C4.5 (Successor of ID3)

The statistical measure used to select attribute (that best splits the dataset in terms of given classes) in ID3 is information gain, whereas in C4.5, the criterion is gain ratio. Both measures have a close relationship with another concept called *entropy*. Indeed, information gain is based on the concept of entropy.

In Section 2.2, we will discuss about entropy by providing practical examples.

2.2 Entropy

The concept of entropy that is used in a machine learning algorithm came from the field of information theory. It is a measure of uncertainty in a random variable. *Shannon entropy* quantifies this uncertainty in terms of *expected value of the information* present in the message.

$$H(X) = \sum_i P(x_i) I(x_i) = -\sum_i P(x_i) \log_b P(x_i)$$

2.2.1 Example

If we toss a *fair* coin, the probability of outcome for head and tail is equal, that is, ½. Hence, in such a maximum uncertain environment, one full bit of information is delivered by each fair toss of coin. In the situation when the coin is unfair or biased, there is less amount of uncertainty or entropy and hence information can be preserved in less number of bits. Tables 2.1 and 2.2 describe outcomes of four tossing experiments each with a fair and unfair coin.

The entropy of dataset D1 can be calculated as follows:

```
Entropy(D1) = -P(H)log₂P(H) + -P(T)log₂P(T)
Entropy(D1) = -2/4log₂(2/4) + -2/4log₂(2/4)
Entropy(D1) = -1/2log₂(1/2) + -1/2log₂(1/2)
Entropy(D1) = -1/2(-1) + -1/2(-1)
Entropy(D1) = 1
```

```
Entropy(D2) = -P(H)log₂P(H)
Entropy(D2) = -4/4log₂(4/4)
Entropy(D2) = 0 that means uncertainty is zero for
unfair coin.
```

Table 2.1   Dataset D1 Describing Outcomes of Four Experiments with Fair Coin

Toss No.	Outcome
1	H
2	T
3	H
4	T

Table 2.2   Dataset D2 Describing Outcomes of Four Experiments with Unfair Coin

Toss No.	Outcome
1	H
2	H
3	H
4	H

2.2.2 Understanding the Concept of Number of Bits

Now consider the example where three possible outcomes exist:

Opinion No. for a Political Issue	Outcome
1	Positive
2	Positive
3	Negative
4	Neutral

```
Entropy(D3) = = -P("Positive")log₂P("Positive")
+ -P("Negative")log₂P("Negative") + -P("Neutral")
log₂P("Neutral")
Entropy(D3) = -2/4log₂(2/4) + -1/4log₂(1/4)
+ -1/4log₂(1/4)
Entropy(D3) = -1/2(-1)+0.25(-2)+ 0.25(-2)
Entropy(D3) = 0.5 + 0.5 + 0.5
Entropy(D3) = 1.5
```

If there were four possible classes, the entropy would be 2 bits with four possible values that are 00, 01, 10, and 11. One can understand the meaning of 1 or 2 bits but what is the meaning of 1.5 bits? We will answer this question in the next paragraph.

Let us apply a simple Huffman coding on the above dataset. We will get the final Huffman code as follows:

Symbol	Code
Positive	0 [1 bit code]
Negative	10 [2 bits code]
Neutral	11 [2 bits code]

Opinion Number for Issue	Outcome	Code	Size in Bits
1	Positive	0	1
2	Positive	0	1
3	Negative	10	2
4	Neutral	11	2
Total number of bits to codify the three outcomes of four opinions			6
Average number of bits to codify the three outcomes of four opinions			6/4 = 1.5 bits

The concept of bits can be used in assessing the potential of compression of a message. In the machine learning world, the same concept is used in another meaning. To build a DT for classification, the target is to find the *attribute* that can reduce the entropy or information content of the whole system by splitting a single dataset into multiple datasets. If we are able to split our dataset in such a manner that resulting datasets are purer (less/no uncertainty or entropy) than machine learning, the job of classification will proceed further. In order to build a DT for classification, the attribute selection is a vital step.

2.3 Attribute Selection Measure

In C4.5, splitinfo is a measure that is used for attribute selection. However, to appreciate the worth of this measure, we will analyze the measure *information gain* that was used in ID3 to select attributes to build a DT.

2.3.1 Information Gain of ID3

We will begin with an example to explain the concept of information gain.

Opinion Number for a Political Issue	Gender	Age	Outcome
1	Male	Below 60	Positive
2	Male	Above 60	Positive
3	Female	Above 60	Negative
4	Female	Below 60	Neutral

Analyzing the above dataset, one will intuitively select "Gender" as the best choice for split. One can derive a simple rule, such as "if Gender is Male then opinion is positive, else it may be negative or neutral." But the question is why the attribute "Gender" was selected? Can we tell the machine to perform a similar operation of selection of "Gender" for such a scenario to split the data. The measure *information gain* performs this job. The attribute that will be selected for splitting purpose will be the one for which the measure of information gain will be the highest.

The formula of information gain for attribute A is given below:

$$\text{Gain}(A) = \text{Info}(D) - \text{Info}_A(D)$$

where:

$$\text{Info}_A(D) = \sum_{j=1}^{v} \frac{|D_j|}{|D|} \times \text{Info}(D_j)$$

Gender	Age	Outcome
Male	Below 60	Positive
Male	Above 60	Positive
Female	Above 60	Negative
Female	Below 60	Neutral

For the above dataset, there are two candidates on the basis of which a split can be made.

The entropy or info (D) of the above dataset is 1.5 (see entropy calculation for dataset D3).

The attribute "Gender" has two possible values that are "Male" and "Female." If "Gender," is selected as an attribute to split the above dataset it will result in two small child datasets. Similarly, the attribute "Age" will split the dataset into two small child datasets for its two values present in the dataset.

Dataset ($D_{\text{gender = male}}$):

Age	Outcome
Below 60	Positive
Above 60	Positive

```
Info(D_gender = male) = 2/2log₂(2/2) = 0
```

$$\text{Info}(D_{\text{gender = male}}) = 2/2\log_2(2/2) = 0$$

Dataset ($D_{\text{gender = Female}}$):

Age	Outcome
Above 60	Negative
Below 60	Neutral

$$\text{Info}(D_{\text{gender = Female}}) = 1/2\log_2(1/2) + 1/2\log_2(1/2) = 1$$
$$\text{Info}_{\text{Gender}}(D) = 2/4(0) + 2/4(1) = 0.5$$
$$\text{Gain}(\text{Gender}) = 1.5 - 0.5 = 1$$

Dataset ($D_{\text{Age = Below60}}$):

Gender	Outcome
Male	Positive
Female	Neutral

```
Entropy(D_Age = Below60) = -1/2log₂(1/2) + -1/2log₂(1/2)
Entropy(D_Age = Below60) = -1/2(-1) + -1/2(-1) =1
```

Dataset ($D_{\text{Age = Above60}}$):

Gender	Outcome
Male	Positive
Female	Negative

```
Entropy(D_Age = Above60) = -1/2log₂(1/2) + -1/2log₂(1/2)
Entropy(D_Age = Above60) = -1/2(-1) + -1/2(-1) =1
Info_Age(D) = 2/4(1) + 2/4(1) =1
Gain(Age) = 1.5 - 1 = 0.5
```

Because Gain (Gender) > Gain (Age), we will select the attribute "Gender." There is no competitor for the attribute "Age" in the next iteration; therefore, it will be selected for further partitioning.

2.3.2 The Problem with Information Gain

The problem with information gain as a measure to select the attribute for partition is that in the quest of *pure* partitions, it can select the attributes that are meaningless from the machine learning point of view. We will demonstrate the problem with the same example but with a new attribute. The new attribute is "Opinion no." for a particular political issue:

Opinion No.	Gender	Age	Outcome
1	Male	Below 60	Positive
2	Male	Above 60	Positive
3	Female	Above 60	Negative
4	Female	Below 60	Neutral

Now we have three candidate attributes that compete to be selected for splitting purpose. Since we have already calculated information gain for "Gender" and "Age" in Section 2.3.1, we will only calculate information gain for the attribute "Opinion no." This will divide the dataset into four pure child datasets.

Dataset ($D_{\text{"Opinion no"} = 1}$):

Gender	Age	Outcome
Male	Below 60	Positive

Dataset ($D_{\text{"Opinion no"} = 2}$):

Gender	Age	Outcome
Male	Above 60	Positive

Dataset ($D_{\text{"Opinion no"} = 3}$):

Gender	Age	Outcome
Female	Above 60	Negative

Dataset ($D_{\text{"Opinion no"}=4}$):

Gender	Age	Outcome
Female	Below 60	Neutral

```
Entropy (D"Opinion no"=1) = 0
Entropy (D"Opinion no"=2) = 0
Entropy (D"Opinion no"=3) = 0
Entropy (D"Opinion no"=4) = 0
Info"Opinion no" (D) = 0
Gain("Opinion no") = 1.5 - 0 = 1.5
```

Thus, Gain ("Opinion no") > Gain("Gender") > Gain("Age"); therefore, the attribute "Opinion no." will be selected as a measure to split the dataset. Since the attribute "Opinion no." has a large number of distinct values, it gives numerous, purest children datasets that are useless for machine learning purpose. This drawback is due to the inherent deficiency in the measure *information gain* that gives preference to the attribute that can divide the parent dataset to datasets with the least amount of entropy. This bias problem leads to a proposal of another measure called *gain ratio* to solve this problem.

2.4 Implementation in MATLAB®

The implementation in MATLAB will help the reader understand the concept in more practical details. Taking the example of a small dataset discussed above, we will explain the problem in the attribute selection measure of "information gain."

The file containing the small dataset is a simple comma separated value (csv) file that will be used as an input source to bring data into the MATLAB environment. The file name is "infogainconcept.csv." The following code describes how the data from the csv file is brought into the MATLAB environment.

```
%test.m

fid = fopen('C:\infogainconcept.csv')';
out = textscan(fid,'%s%s%s%s','delimiter',',');
fclose(fid);
num_featureswithclass = size(out,2);
tot_rec = size(out{size(out,2)},1)-1;
for i = 1:tot_rec
```

```
        yy{i} = out{num_featureswithclass}{i+1};
end
for i = 1: num_featureswithclass
    xx{i} = out{i};
end
[a,b] = infogain(xx,yy);
```

The function "infogain" (corresponding file, infogain.m) takes the input of xx (consisting of information about all input features) and yy (information about class feature). In order to calculate the information gain of each feature, the entropy of the dataset should be calculated. The following piece of code from the function "infogain" calculates the entropy of the dataset.

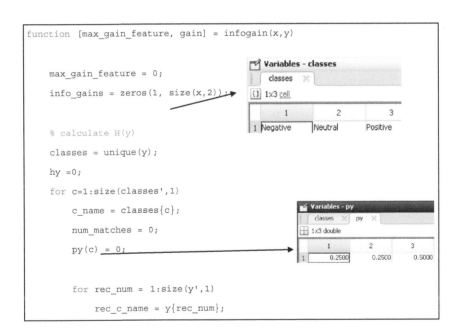

In order to calculate the infogain of each feature, the following code snippet will do the job.

```
% function [max_gain_feature, gain] = infogain(x,y) ... continued

p=zeros(size(x,2),10,size(classes',1)); %
    % iterate over all features (columns)
    for col=1:size(x,2)
        feature_with_header{col} = x{col};
        num_rec = size(y',1);
        for i = 1:num_rec
            feature_without_header{i} = feature_with_header{col}{i+1}; %
for example [1 1 1 1]
        end
        features = unique(feature_without_header); %find unique values of each
feature
        % calculate entropy
        hyx = 0;
        for c=1:size(features',1)
            f_name = features{c};
            num_matches_feature = 0;
            num_matches_feature_class = 0;
            py(c) = 0;
            for rec_num = 1:size(feature_without_header',1)
                rec_c_name = feature_without_header{rec_num};
                if strcmp(rec_c_name, f_name) == 1
                    num_matches_feature = num_matches_feature + 1;
                    for c1=1:size(classes',1)
                        c_name = classes{c1};
                        if (strcmp(y{rec_num},c_name)==1) && (col ~= size(x,2))
                            p(col,c,c1) = p(col,c,c1)+1;
                        end
                    end
                end
            end
        end
    end
    q = sum(p,3); %Adding based on class variable
    jprobtot = zeros(size(x,2),10);
    for i = 1:size(classes',1)
        j{i} = p(:,:,i)./q;
        j{i}(isnan(j{i})) = 0;
        jprob{i} = j{i} .* log2(j{i});
        jprob{i}(isnan(jprob{i})) = 0;
    end
    probmat = cell2mat(jprob);
    num_col = size(probmat,2);
    step_size = num_col/size(classes',1);
    for i = 1: step_size:num_col
        jprobtot = jprobtot + probmat(:,i:i+step_size-1);
    end
    qtot = q/sum(q(1,:));
    EntropyPerFeature = jprobtot .* qtot;
    EntropyFinal = -1 * sum(EntropyPerFeature,2);
    info_gains = hy - EntropyFinal;
    info_gains = info_gains(1:size(info_gains)-1);

    [gain, max_gain_feature] = max(info_gains);
```

info_gains	
3x1 double	
	1
1	1.5000
2	1
3	0.5000

The above code snippet calculates the entropy of each feature and then calculates the information gain by each feature. For three features, the following values of information gain were calculated:

1. Num: 1.5000
2. Gender: 1
3. Age: 0.500

Since the information gain of the attribute "Num" is highest, it will be used as a splitting point. In the above scenario, the attribute selection on the basis of the "information gain" criterion is useless and therefore it should be improved. "Gain Ratio," as an improved attribute selection criterion will be discussed next.

2.4.1 Gain Ratio *of C4.5*

In order to solve the bias problem of the measure of information gain, another measure called *gain ratio* was proposed. The idea behind the new measure is very simple and intuitive. Just penalize those attributes that make many splits. Hence, attributes like "Opinion no." will be penalized for *too much* partitioning. The extent of partitioning is calculated by *SplitInfo*. This normalizes the *information gain*, resulting in the calculation of gain ratio.

$$\text{SplitInfo}_A(D) = -\sum_{j=1}^{v} \frac{|D_j|}{|D|} \times \log_2\left(\frac{|D_j|}{|D|}\right)$$

$$\text{Gain ratio}(A) = \frac{\text{Gain}(A)}{\text{SplitInfo}(A)}$$

We will again use our small dataset to illustrate the concept:

Opinion No.	Gender	Age	Outcome
1	Male	Below 60	Positive
2	Male	Above 60	Positive
3	Female	Above 60	Negative
4	Female	Below 60	Neutral

```
SplitInfo"opinion no" (D) = -1/4*log(1/4) +
-1/4*log(1/4)+ -1/4*log(1/4)+ -1/4*log(1/4)
SplitInfo"opinion no" (D) = 0.5 + 0.5 + 0.5 + 0.5 = 2
SplitInfo Age (D) = -2/4*log(2/4) + -2/4*log(2/4)
=0.5 + 0.5 = 1
SplitInfo Gender (D) = -2/4*log(2/4) + -2/4*log(2/4)
=0.5 + 0.5 = 1
GainRatio("Opinion No") = gain("Opinion No")/
splitinfo("Opinion No")
GainRatio("Opinion No") =1.5/2 = 0.75
Similarly GainRatio("Gender") = gain(Gender)/
Splitinfo(Gender) = 1/1 = 1
GainRatio(Age) = gain(Age)/Splitinfo(Age) = 0.5/1 =
0.5
Hence GainRatio(Gender) > GainRatio("Opinion no") >
GainRatio("Age")
```

Therefore, "Gender" will be selected as an attribute to split the dataset in the first iteration.

In the next iteration, attributes "Age" and "Opinion no." will compete to be selected as an attribute for splitting the datasets from the first iteration. One of the two datasets, that is, Dataset ($D_{gender = male}$), has 0 entropy; therefore, there is no need for a further split. The other dataset ($D_{gender = Female}$) has a nonzero entropy, so we need to select an attribute. The two competing attributes have same values for information gain and SplitInfo and, hence, the measure "Gain Ratio" is also the same. When such a situation arises, a decision can be taken arbitrarily.

From the above example it is clear that the value of "SplitInfo" increases as the number of partitions increases. Consider three scenarios with evenly distributed partitions.

- 10 possible partitions with even distribution
  SplitInfo = $-10*(1/10*\log2(1/10)) = 3.32$
- 100 possible partitions with even distribution
  SplitInfo = $-100*(1/100*\log2(1/100)) = 6.64$

■ 1000 possible partitions with even distribution
SplitInfo = −1000*(1/1000*log2(1/1000)) = 9.97

It can be seen that as the number of partitions increases, the value of the SplitInfo goes up, which in turn reduces the value of Gain Ratio for a particular attribute.

2.4.2 *Implementation in MATLAB*

We have to make small changes to the function of "infogain." The new function name is "gaininfo." The file "gaininfo.m" serves the purpose for calculation of "gaininfo" of each feature and then returning the one with the maximum number of value of "gaininfo." The following code snippet describes the "gaininfo" calculation.

```
% To calculate split info of each attribute
     r = q; [row_r,col_r] = size(r);
     split_count = zeros(row_r,1)
     splitinfo = zeros(row_r,1)
     for i = 1:row_r
              for jj = 1:col_r
              if r(i,jj) ~= 0
                   split_count(i) = split_Count(i)+
                   r(i,jj);
              end
              end
              for jj = 1:col_r
              if r(i,jj) ~= 0
                   r(i,jj) = r(i,jj)/
                   split_count(i);
                   splitinfo(i) = splitinfo(i) +
              r(i,jj)* log2(r(i,jj));
              end
              end
     splitinfo(i) = -1.* splitinfo(i);
     end
```

After calculating the "SplitInfo" of each attribute, the "Gain Ratio" is calculated using the following code.

```
Gain_ratio = info_gains./splitinfo
[gain, max_gain_feature] = max(Gain_ratio);
```

Now the criterion of the selection of an attribute for making a DT is improved.

The two functions "infogain" and "gainratio" can be used to develop a DT easily.

References

1. Jiawei, H. and Kamber, M. *Data Mining: Concepts and Techniques*, 2nd Ed. Burlington, MA: Elsevier, 2006, p. 772.
2. Quinlan, J.R. Induction of decision trees, *Machine Learning*, vol.1, 81–106, 1986.

Chapter 3

Rule-Based Classifiers

3.1 Introduction to Rule-Based Classifiers

Rule-based classifiers use a set of IF-THEN rules for classification. We can express a rule in the following form:

```
IF condition THEN decision
Let us consider a rule R,
R: IF (humidity = "high") and (outlook = "sunny")
THEN play = "no"
```

Following are some terminologies that are used to describe the different aspects of the IF-THEN rules.

1. The IF part of the rule is called *rule antecedent*. In Rule **R**, it is (humidity = "high") and (outlook = "sunny").
2. The THEN part of the rule is called *rule consequent*. In Rule R, it is play = "no."
3. The rule antecedent consists of one or more attribute tests combined together with the logical AND operator.
4. The rule consequent consists of a decision of the classifier if the conditions mentioned in the antecedent hold for a particular example or tuple.

3.2 Sequential Covering Algorithm

A generic rule-learning classification task for a two-class problem is to classify the instance (or example or tuple) into *positive* or *negative* classes. For the scenario with more than two classes, the job will be broken into more steps. For example, if there are three classes namely c1, c2, and c3, then we can treat c1 (or c2 or c3) as positive class and the combination of c2 and c3 (or c1 and c3 or c1 and c2) as negative class. After the discovery of rules for c1 (or c2 or c3), we can perform a binary classification by assigning c2 (or c1 or c3) and c3 (or c1 or c2) to the positive and negative classes. For more than three classes, a similar strategy can be adopted with different sets of combination.

Sequential covering algorithm [1] is popular for rule learning. The words *sequential* and *covering* in the name of the algorithm describe a lot about the nature of the algorithm.

- It is *covering* in the sense that it develops a cover of rules for the set of examples (or tuples) for each class
- It is *sequential covering* because it learns one rule at a time and repeats this process to gradually cover the full set of examples (or tuples) for each class

Ideally, a ruleset for a positive class should cover only the examples of the positive class and should not include examples from the negative class, but in certain situations, a compromise is made to accept the *imperfect* rules. The algorithm of sequential covering (discussed in Section 3.3) uses the *Learn-One-Rule* procedure that finds available instances of the best rule for the current class.

3.3 Algorithm

The steps for the sequential covering algorithm are given below:

```
Input: A labeled Data set
Output: A Set of IF-THEN rules.
Method:
```

1. *Start with an empty **Ruleset named RS**.*
2. For each class **c**, perform the following steps:
   - *Repeat*
      i. *Use **Learn-One-Rule** to find the best rule **R**.*
      ii. *Remove examples/tuples covered by **R**.*
   - *Until the terminating condition is satisfied. //e.g. no tuples left or quality of rule is below specific threshold.*
   - *Add **R** to **RS**.*
3. *Return: **RS**.*

3.4 Visualization

To discover the rules for classification of sheep and goats, the sequential covering problem takes the target class of sheep as the positive class and then sequentially grows the rule until it covers all positive cases. Figure 3.1 depicts the process in four stages.

The rule-based classifiers are *not black boxes* for us like neural networks. They are interpretable just like decision trees. However, they often have lower coverage and, hence, there can be tuples that cannot be classified using the classifier developed by the training set using the rule-based induction.

3.5 Ripper

Ripper (repeated incremental pruning to produce error reduction) is one of the popular sequential covering algorithms. It is based on the technique called *reduced error pruning* (REP) which is also used in decision tree algorithms. Rule-based algorithms use REP by splitting their training data into a growing set and a pruning set [2]. The initial ruleset is formed using the growing set and the ruleset is simplified

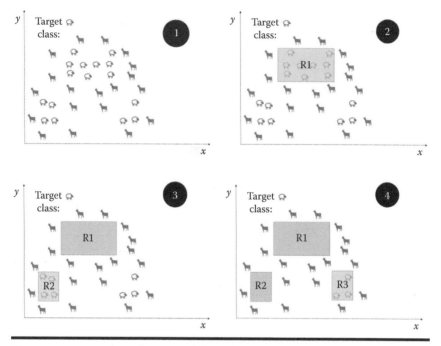

Figure 3.1 Process of sequential covering algorithm to grow rules to cover "Sheep" class.

by using the pruning operation that will yield the greatest reduction of error on the pruning set. The pruning operation can be deletion of any single condition or any single rule. The pruning stops when further application of the pruning operator results in an increased error on the pruning set. In order to explain how the Ripper works, we will use JRip, an implementation of Ripper in Weka.

3.5.1 Algorithm

The algorithm of JRip is described in Wikibooks* and is given as follows:

```
Initialize RS = {} // initially there is nothing in
RuleSet
```

* https://en.wikibooks.org/wiki/Data_Mining_Algorithms_In_R/Classification/JRip.

For each *class from the less prevalent one to the more frequent one*, DO:

1. *Building stage:* Repeat steps 1.1 and 1.2 until the *description length* (DL) of the ruleset and examples is 64 bits greater than the smallest DL met so far, or there are no positive examples, or the error rate >= 50%.

    Step:1.1 Grow phase: Grow one rule by greedily adding antecedents (or conditions) to the rule until the rule is perfect (i.e., 100% accurate). The procedure tries every possible value of each attribute and selects the condition with the highest information gain: $p(\log(p/t) - \log(P/T))$.

    Step:1.2. Prune phase: Incrementally, prune each rule and allow the pruning of any final sequences of the antecedents. The pruning metric is $(p - n)/(p + n)$ – but it is actually $2p/(p + n) - 1$. So, in this implementation, we simply use $p/(p + n)$ (actually $(p + 1)/(p + n + 2)$; thus, if $p + n$ is 0, it is 0.5).

2. *Optimization stage:* After generating the initial ruleset $\{R_i\}$, generate and prune two variants of each rule R_i from the randomized data using procedures 1.1 and 1.2.

    After the generation of the ruleset,

    – For each rule r in the ruleset **R**
        • Consider 2 alternative rules:
            ■ Replacement rule (r^*): grow new rule from scratch
            ■ Revised rule (r'): add conjuncts to extend the rule r
        • Compare the ruleset for r against the ruleset for r^* and r'
        • Choose the ruleset that minimizes the minimum description length (MDL) principle

But one variant is generated from an empty rule, whereas the other is generated by greedily adding antecedents to the original rule. Moreover, the pruning metric used here is $(TP + TN)/(P + N)$. Then the smallest possible DL for each variant and the original rule is computed. The variant with the minimal DL is selected as the final representative of R_i in the ruleset. After all the rules in $\{R_i\}$ have been examined and if there are still residual positives, more rules are generated based on the residual positives using the building stage again.

3. Delete the rules from the ruleset that would increase the DL of the whole ruleset if it were in it and add the resultant ruleset to RS.

ENDDO

Because of certain reasons, the implementation of JRip is slightly different from the original Ripper program.

To demonstrate how the Ripper works, we will present our example in three scenarios.

1. No pruning is used (just basic rule growing)
2. Incremental pruning is used
3. Optimization is performed

3.5.2 Understanding Rule Growing Process

To understand the phenomenon of rule growing, we will use the following benchmark dataset:

Serial No.	Outlook	Temperature	Humidity	Windy	Play
1	Sunny	Hot	High	False	No
2	Sunny	Hot	High	True	No
3	Overcast	Hot	High	False	Yes

(Continued)

Serial No.	Outlook	Temperature	Humidity	Windy	Play
4	Rainy	Mild	High	False	Yes
5	Rainy	Cool	Normal	False	Yes
6	Rainy	Cool	Normal	True	No
7	Overcast	Cool	Normal	True	Yes
8	Sunny	Mild	High	False	No
9	Sunny	Cool	Normal	False	Yes
10	Rainy	Mild	Normal	False	Yes
11	Sunny	Mild	Normal	True	Yes
12	Overcast	Mild	High	True	Yes
13	Overcast	Hot	Normal	False	Yes
14	Rainy	Mild	High	True	No

As can be seen from the above table, there are two classes in the dataset. The number of tuples belonging to class "No" is 5, whereas the number of tuples belonging to class "Yes" are 9. Since in this section we will discuss the rule growth process without pruning, there will be no division of data in folds and hence all 14 tuples will be used for the growth.

Since the "No" class is a *less prevalent* class, the Ripper will try to grow rules to cover the instances belonging to this class. The class "Yes" will be treated as a default class. For a multiclass problem, the classes are ordered according to their prevalence in the dataset and then for each class, except the most prevalent class (default class), the rules are grown and pruned. For example, if there are 14 tuples with 3, 4, and 7 tuples belonging to Class C1, C2, and C3, respectively, then C3 will be the default class. First, the Ripper algorithm will try to find the rules covering the tuples belonging to C1. In the next step, it will try to find the rules that can cover tuples belonging to Class C2.

We return to our example problem where for four attributes, we had 10 possibilities as can be seen in following summary:

Outlook (Sunny, Overcast, Rainy)
Temperature (Hot, Mild, Cool)
Humidity (High, Normal)
Windy (False, True)

The algorithm will start to grow a rule for every possible value of an attribute. Since the attribute "Outlook" has three possible values, it can produce three possible rules. The number of instances that will be covered by the antecedent of these rules and their accuracies are also mentioned with each rule:

Rule	Antecedent Coverage	True Positive (Number of Tuples for Which Rule Is Satisfied)	Accuracy
Outlook = Sunny -> Play = No	5	3	3/5 = 0.6
Outlook = Overcast -> Play = No	4	0	0
Outlook = Rainy -> Play = No	4	2	2/4 = 0.5

The Ripper algorithm uses "Information Gain" as a measure to select one rule from the competing ones.

$$\text{Gain}(R',R) = s \cdot \left(\log_2 \frac{N'_+}{N'} - \log_2 \frac{N_+}{N} \right)$$

where:
R is the original rule
R' is the candidate rule after adding a condition

N (N') is the number of instances that are covered by R (R')

N_+ (N'_+) is the number of true positives in R (R')

s is the number of true positives in R and R' (after adding the condition)

In the beginning when no rule is selected, the following holds true for R:

$$N_+ = 5; N = 14$$

If R' is the first rule given in the above table that is "Outlook = Sunny –> Play = No," then the following values hold true for R':

$$N'_+ = 3; N' = 5$$

The value of s will be 3. Hence, Gain ("Outlook = Sunny –> Play = No," R) = 3 ($\log_2 3/5 - \log_2 5/14$). Similarly, Gain ("Outlook = Overcast –> Play = No," R) = 0. And Gain ("Outlook = Rainy –> Play = No," R) = 2 ($\log_2 2/5 - \log_2 5/14$). Since Gain for the rule "Outlook = Sunny –> Play = No" is highest, this rule will be selected for further expansion.

Applying the same procedure will result in obtaining the following rules for further expansion:

Temperature = Hot –> Play = No
Humidity = High –> Play = No
Windy = True –> Play = No

Since the information gain of the rule "Humidity = High –> Play = No" is the highest among four rules, we will add a condition to this rule to get the improved rule. The antecedent of the rule covers 7 tuples; therefore, further checking will be made on these 7 tuples:

Serial No.	Outlook	Temperature	Humidity	Windy	Play
1	Sunny	Hot	High	False	No
2	Sunny	Hot	High	True	No
3	Overcast	Hot	High	False	Yes
4	Rainy	Mild	High	False	Yes
5	Sunny	Mild	High	False	No
6	Overcast	Mild	High	True	Yes
7	Rainy	Mild	High	True	No

The three rules ("Outlook = Sunny –> Play = No," "Temperature = Hot –> Play = No," "Windy = True –> Play = No") are checked.

The information gain of the rule "Outlook = Sunny –> Play = No" is the highest among the three contending rules for the dataset with 7 tuples; therefore, it will be selected as a condition to be added to the previously derived rule. The calculation of information gain of the rule "Outlook = Sunny –> Play = No" is given below:

Since the rule R is "humidity = high –> play = no":

$$N_+ = 4; \ N = 7$$

R' is the rule "Outlook = Sunny –> Play = No," following values hold true for R':

$$N'_+ = 3$$

$$N' = 3$$

The value of s will be 3 again. Hence, Gain (Humidity = High) and (Outlook = Sunny) => Play = No, "Humidity = High –> Play = No") = 3 ($\log_2 3/3 - \log_2 4/7$). The resulting rule will be (Humidity = High) and (Outlook = Sunny) => Play = No:

Serial No.	Outlook	Temperature	Humidity	Windy	Play
1	Sunny	Hot	High	False	No
2	Sunny	Hot	High	True	No
3	Overcast	Hot	High	False	Yes
4	Rainy	Mild	High	False	Yes
5	Sunny	Mild	High	False	No
6	Overcast	Mild	High	True	Yes
7	Rainy	Mild	High	True	No

The rule has 100% accuracy, and thus after the growth of the rule, we are left with 11 tuples containing 9 yes and 2 No:

Serial No.	Outlook	Temperature	Humidity	Windy	Play
1	Overcast	Hot	High	False	Yes
2	Rainy	Mild	High	False	Yes
3	Rainy	Cool	Normal	False	Yes
4	Rainy	Cool	Normal	True	No
5	Overcast	Cool	Normal	True	Yes
6	Sunny	Cool	Normal	False	Yes
7	Rainy	Mild	Normal	False	Yes
8	Sunny	Mild	Normal	True	Yes
9	Overcast	Mild	High	True	Yes
10	Overcast	Hot	Normal	False	Yes
11	Rainy	Mild	High	True	No

The same procedures will be applied to the grow rule for the remaining two "No" tuples.

The four rules "Outlook = Rainy –> Play = No," "Temperature = Cool –> Play = No," "Humidity = High –> Play = No," and "Windy = True –> Play = No" compete. The information gain for the two rules "Outlook = Rainy –> Play = No" and "Windy = True–> Play = No" is equal and highest. We will select "Outlook = Rainy –> Play = No" as a rule to be further grown.

The selected rule will lead to the selection of 5 tuples on which the antecedent of the selected rule holds true. The dataset is given below:

Serial No.	Outlook	Temperature	Humidity	Windy	Play
1	Rainy	Mild	High	False	Yes
2	Rainy	Cool	Normal	False	Yes
3	Rainy	Cool	Normal	True	No
4	Rainy	Mild	Normal	False	Yes
5	Rainy	Mild	High	True	No

The three rules "Temperature = Cool –> Play = No," "Humidity = High –> Play = No," and "Windy = True –> Play = No" will compete. Since the information gain of the rule "Windy = True –> Play = No" is the highest, the condition will be appended to the original rule. The resulting rule will be

```
(Outlook = Rainy) and (Windy = True) => Play = No
```

Thus, the final ruleset will consist of three rules (including default rule).

Rule 1: (Humidity = High) and (Outlook = Sunny) => Play = No

Rule 2: (Outlook = Rainy) and (Windy = True) => Play = No

Rule 3: => Play = Yes

Rule 3 should be read as "All the tuples that do not satisfy Rule 1 or Rule 2 are the one who has the class of 'Yes.'"

3.5.3 *Information Gain*

We have discussed that information gain is the measure to select a rule from a set of competing rules. There is another measure that can possibly be used to perform this job, which is accuracy. The question is why we do not use accuracy as a measure to be used for rule selection. In the next few lines, we will try to answer this question. Suppose that we have the following dataset:

Outlook	Temperature	Humidity	Windy	Play
Sunny	Mild	High	False	No
Sunny	Mild	Normal	True	Yes
Sunny	Hot	High	True	No
Rainy	Mild	High	False	Yes
Rainy	Mild	Normal	False	Yes
Rainy	Cool	Normal	False	Yes

Now we will compare the accuracy of the two competing rules:

Rule	Antecedent Coverage	True Positive (Number of Tuples for Which Rule Is Satisfied)	Accuracy
Outlook = Sunny –> Play = No	3	2	2/3 = 0.666
Temperature = Hot –> Play = No	1	1	1

If we compare accuracy, we will select the rule "Temperature = Hot –> Play = No" for better performance in terms of the said measure. Now we will compare the two rules with respect to information gain. The formula for information gain is again given below:

$$\text{Gain}(R', R) = s \cdot \left(\log_2 \frac{N'_+}{N'} - \log_2 \frac{N_+}{N} \right)$$

Rule No.	Rule	N_+	N	N'_+	N'	S(True Positive)	Accuracy	Information Gain
1	Outlook = Sunny -> Play = No	2	6	2	3	2	2/3 = 0.666	2
2	Temperature = Hot -> Play = No	2	6	1	1	1	1	1.584963

Even though the accuracy of rule 2 is higher than the accuracy of rule 1, the information gain take cares of another important variable, that is the number of true positives covered by the rule (the variable s), and since the number of true positives covered by rule 1 is higher than that of rule 2, the information gain gives different preferences for the rule selection.

3.5.4 Pruning

To manage the over-fitting problem, pruning is used. In order to perform the pruning, the data is divided into two parts. One part is used to grow the rule, and the other part is used to prune the grown rule. In order to demonstrate the working of pruning, we have slightly modified the dataset. The class of the following tuple is replaced with "No."

Sunny	Cool	Normal	False	Yes/No

Two datasets for growing and pruning rules, respectively, are given below:

Dataset for growing rule:

Outlook	Temperature	Humidity	Windy	Play
Sunny	Hot	High	False	No
Sunny	Hot	High	True	No
Rainy	Mild	High	False	Yes
Rainy	Cool	Normal	False	Yes
Rainy	Cool	Normal	True	No
Sunny	Cool	Normal	False	No
Rainy	Mild	Normal	False	Yes
Sunny	Mild	Normal	True	Yes
Overcast	Mild	High	True	Yes

Dataset for pruning grown rules:

Outlook	Temperature	Humidity	Windy	Play
Overcast	Cool	Normal	True	Yes
Sunny	Mild	High	False	No
Rainy	Mild	High	True	No
Overcast	Hot	High	False	Yes
Overcast	Hot	Normal	False	Yes

In the process of growing a rule, the rule "Outlook = Sunny –> Play = No" will win from other competing rules. The addition of the condition to this rule will also lead to the growth of a few more rules and the same process that is described above will lead to the selection of the rule "Outlook = Sunny and Temperature = Hot –> Play = No."

After the process of the growth of rules, the grown rule is tested on the data that is held for the pruning process:

Outlook	Temperature	Humidity	Windy	Play	Outlook = Sunny and Temperature = Hot –> Play = No	Outlook = Sunny –> Play = No
Overcast	Cool	Normal	True	Yes		
Sunny	Mild	High	False	No		No
Rainy	Mild	High	True	No		
Overcast	Hot	High	False	Yes		
Overcast	Hot	Normal	False	Yes		

The pruning metric is $(p - n)/(p + n)$. The same metric can be written as $2p/(p + n) - 1$. Discarding constants, we can take $(p/p + n)$ as an equivalent measure.

For the full grown rule "Outlook = Sunny and Temperature = Hot –> Play = No," the value of $p = 0$, $n = 0$. In order to avoid 0/0 problem, we will use the Laplace correction. The metric will become $(p + 1)/(p + n + 2)$.

Hence, for the rule "Outlook = Sunny and Temperature = Hot –> Play = No," the pruning metric will have the value of $(0 + 1)/(0 + 0 + 2) = 0.5$. For the pruned rule with only one condition that is "Outlook = Sunny –> Play = No," $p = 1$ and $n = 0$; therefore, the pruning metric will have the value of $(1 + 1)/(1 + 0 + 2) = 2/3 = 0.666$.

Since the reduced rule has a better value for the pruning metric, the full grown rule will be discarded and the reduced rule with a single condition will be retained to make the ruleset.

3.5.5 Optimization

In this section, we will explain the whole process of the Ripper algorithm along with optimization.

The following datasets are meant for growing and pruning rules.

Dataset for growing rules:

Outlook	Temperature	Humidity	Windy	Play
Sunny	Hot	High	False	No
Sunny	Hot	High	True	No
Overcast	Hot	High	False	Yes
Rainy	Mild	High	False	Yes
Rainy	Cool	Normal	False	Yes
Rainy	Cool	Normal	True	No
Overcast	Cool	Normal	True	Yes
Sunny	Mild	High	False	No
Rainy	Mild	Normal	False	Yes
Overcast	Hot	Normal	False	Yes
Rainy	Mild	High	True	No

Dataset for pruning rules:

Outlook	Temperature	Humidity	Windy	Play
Sunny	Cool	Normal	False	Yes
Sunny	Mild	Normal	True	Yes
Overcast	Mild	High	True	Yes

We will get the following rule:

```
(Outlook = Sunny) -> Play = No
```

The performance of the above rule is very good in the dataset dedicated for the growing rule, but it was unable to classify the two related tuples correctly in the dataset dedicated for the pruning rules. However, we will retain this rule that covers the 5 tuples in the two datasets. The other rule that will grow in the process and will survive the pruning stage will be

```
(Windy = True) and (Outlook = Rainy) => Play = No
```

Thus, our ruleset will consist of two rules:

R1: (Outlook = Sunny) -> Play = No
R2: (Windy = True) and (Outlook = Rainy) => Play = No

Now the process of optimization will start. In this process, we will generate two variants of each rule in our ruleset that is generated in the previous steps. One variant, the *replacement rule* (R*) will be grown from scratch. The other variant *revised rule* (R') will be generated by greedily adding antecedents to the original rule. For our example, the following steps will be performed:

- Generation of R1* and R1' for rule 1
- Calculation of MDL for two variants of R1: MDL (R1*, R2), MDL (R1', R2)
- Comparison of three MDL values that are MDL (R1, R2), MDL (R1*, R2), and MDL (R1', R2)
- Choose the ruleset with the minimum MDL value. We will call the first rule R1+ in the new ruleset
- Generation of R2* and R2' for rule 2
- Calculation of MDL for two variants of R2: MDL (R1+, R2*) and MDL (R1+, R2')
- Comparison of the three MDL values that are MDL (R1+, R2), MDL (R1+, R2*), and MDL (R1+, R2')
- Choose the ruleset with the minimum MDL value

We will not discuss the issue of MDL in this book. Interested readers are encouraged to read the related research papers in order to grasp the concepts.

To generate two variants of each rule, the optimization process will generate its own dataset for the growing and the pruning rule. For example, the dataset for the optimization stage for the growing and pruning rules are given below.

Dataset for the growing rule in the optimization stage:

Outlook	Temperature	Humidity	Windy	Play
Sunny	Mild	High	False	No
Overcast	Hot	High	False	Yes
Overcast	Mild	High	True	Yes
Rainy	Cool	Normal	True	No
Rainy	Mild	High	False	Yes
Sunny	Mild	Normal	True	Yes
Sunny	Hot	High	True	No
Overcast	Hot	Normal	False	Yes
Sunny	Hot	High	False	No
Rainy	Mild	Normal	False	Yes
Rainy	Mild	High	True	No

Dataset for the pruning rules in the optimization stage:

Outlook	Temperature	Humidity	Windy	Play
Sunny	Cool	Normal	False	Yes
Overcast	Cool	Normal	True	Yes
Rainy	Cool	Normal	False	Yes

When we will grow the replacement $R1^*$ for $R1$, we will get the following replacement rule after the growing and pruning steps.

```
R1* = (Outlook = Sunny) and (Humidity =
High) => Play = No
```

But this is also the best option for revising the rule $R1$; therefore,

```
R1'= (Outlook = Sunny) and (Humidity = High) =>
Play = No
```

Now the algorithm will calculate the MDL for the three rulesets. Since the basic change will only be for the rules related to $R1$ in this step, it will be enough to compare the DL of the three rules. Calculation of the DL of a rule is out of the scope of this book. The DL of $R1^*$ and $R1'$ is lesser than $R1$, therefore they will replace R1 in the ruleset. Similar procedures will be done for $R2$. In this way, the optimization will further ensure that the resulting ruleset classifies the dataset in an improved manner.

References

1. Jiawei, H. and Kamber, M. *Data Mining: Concepts and Techniques*, 2nd Ed. Burlington, MA: Elsevier, 2006, p. 322.
2. Cohen, W. W. Fast effective rule induction, *Proceedings of the 12th International Conference on Machine Learning*, Lake Tahoe, CA, 1995.

Chapter 4

Naïve Bayesian Classification

4.1 Introduction

Naïve Bayesian classifiers [1] are simple probabilistic classifiers with their foundation on application of Bayes' theorem with the assumption of strong (naïve) independence among the features. The following equation [2] states Bayes' theorem in mathematical terms:

$$P(A \mid B) = \frac{P(A)P(B \mid A)}{P(B)}$$

where:
  A and B are events
  $P(A)$ and $P(B)$ are the prior probabilities of A and B without
      regard to each other
  $P(A \mid B)$, also called *posterior probability*, is the probability of
      observing event A given that B is true
  $P(B \mid A)$, also called *likelihood*, is the probability of observing
      event B given that A is true

Suppose that vector $\mathbf{X} = (x_1, x_2, \ldots x_n)$ is an instance (with n independent features) to be classified and c_j denotes one of K classes, then using Bayes' theorem we can calculate the posterior probability, $P(c_j | X)$, from $P(c_j)$, $P(X)$, and $P(X | c_j)$. The naïve Bayesian classifier makes a simplistic (naïve) assumption called *class conditional independence* that the effect of the value of a predictor (x_i) on a given class c_j is independent of the values of other predictors.

Without going into mathematical details, for each of the K classes, the calculation of $P(c_j | X)$ for $j = 1$ to K is performed. The instance X will be assigned to class c_k, if and only if

$$P(c_k | \mathbf{X}) > P(c_j | \mathbf{X}) \quad \text{for} \quad 1 \leq j \leq K, j \neq k$$

The idea will be further clear when the example classification using the naïve Bayesian classifier will be discussed along with the implementation of MATLAB®.

4.2 Example

To demonstrate the concept of the naïve Bayesian classifier, we will again use the following dataset:

Outlook	Temperature	Humidity	Windy	Play
Sunny	Hot	High	False	No
Sunny	Hot	High	True	No
Overcast	Hot	High	False	Yes
Rainy	Mild	High	False	Yes
Rainy	Cool	Normal	False	Yes
Rainy	Cool	Normal	True	No
Overcast	Cool	Normal	True	Yes
Sunny	Mild	High	False	No

(Continued)

Outlook	Temperature	Humidity	Windy	Play
Sunny	Cool	Normal	False	Yes
Rainy	Mild	Normal	False	Yes
Sunny	Mild	Normal	True	Yes
Overcast	Mild	High	True	Yes
Overcast	Hot	Normal	False	Yes
Rainy	Mild	High	True	No

4.3 Prior Probability

Our task is to predict using different features whether tennis will be played or not. Since there are almost twice as many examples of "Play=yes" (9 examples) as compared to examples of "Play=No" (5 examples), it is reasonable to believe that a new unobserved case is almost twice as likely to have class of "Yes" as compared to "No." In the Bayesian paradigm, this belief, based on previous experience, is known as the *prior probability*.

Since there are 14 available examples, 9 of which are Yes and 5 are No, our prior probabilities for class membership are as follows:

```
Prior Probability P(Play = Yes) = 9 / 14
Prior Probability P(Play = No) = 5 / 14
```

4.4 Likelihood

Let X be the new example for which we want to predict that tennis is going to be played or not. We can assume that the more (Play = Yes) (or No) examples are closer to X, the more *likely* that the new cases belong to (Play = Yes) (or No).

Let X = (Outlook = Overcast, Temperature = Mild, Humidity = Normal, Windy = False), then we have to compute the conditional probabilities that are given as underlined text in the following table:

Outlook	
$P(\text{Sunny}\mid \text{Play} = \text{Yes}) = 2/9$	$P(\text{Sunny}\mid \text{Play} = \text{No}) = 3/5$
$\underline{P(\text{Overcast}\mid \text{Play} = \text{Yes}) = 4/9}$	$\underline{P(\text{Overcast}\mid \text{Play} = \text{No}) = 0}$
$P(\text{Rain}\mid \text{Play} = \text{Yes}) = 3/9$	$P(\text{Rain}\mid \text{Play} = \text{No}) = 2/5$
Temperature	
$P(\text{Hot}\mid \text{Play} = \text{Yes}) = 2/9$	$P(\text{Hot}\mid \text{Play} = \text{No}) = 2/5$
$\underline{P(\text{Mild}\mid \text{Play} = \text{Yes}) = 4/9}$	$\underline{P(\text{Mild}\mid \text{Play} = \text{No}) = 2/5}$
$P(\text{Cool}\mid \text{Play} = \text{Yes}) = 3/9$	$P(\text{Cool}\mid \text{Play} = \text{No}) = 1/5$
Humidity	
$P(\text{High}\mid \text{Play} = \text{Yes}) = 3/9$	$P(\text{High}\mid \text{Play} = \text{No}) = 4/5$
$\underline{P(\text{Normal}\mid \text{Play} = \text{Yes}) = 6/9}$	$\underline{P(\text{Normal}\mid \text{Play} = \text{No}) = 1/5}$
Windy	
$P(\text{True}\mid \text{Play} = \text{Yes}) = 3/9$	$P(\text{True}\mid \text{Play} = \text{No}) = 3/5$
$\underline{P(\text{False}\mid \text{Play} = \text{Yes}) = 6/9}$	$\underline{P(\text{False}\mid \text{Play} = \text{No}) = 2/5}$

Using the above probabilities, we can obtain the two probabilities of the likelihood of X belonging to any of the two classes:

1. $P(X/\text{Play} = \text{Yes})$
2. $P(X/\text{Play} = \text{No})$

The two probabilities can be obtained by the following calculations:

$$P(X/\text{Play} = \text{Yes}) = P(\text{Outlook} = \text{overcast}\mid \text{play} = \text{Yes}) \times P(\text{Temperature} = \text{mild}\mid \text{play} = \text{Yes}) \times P(\text{Humidity} = \text{normal}\mid \text{play} = \text{Yes}) \times P(\text{Windy} = \text{false}\mid \text{play} = \text{Yes})$$

$$P(X/\text{Play} = \text{No}) = P(\text{Outlook} = \text{overcast}\mid \text{play} = \text{No}) \times P(\text{Temperature} = \text{mild}\mid \text{play} = \text{No}) \times P(\text{Humidity} = \text{normal}\mid \text{play} = \text{No}) \times P(\text{Windy} = \text{false}\mid \text{play} = \text{No})$$

4.5 Laplace Estimator

One of the evident problems in calculating $P(X/Play = No)$ is the presence of the value of zero for the conditional probability $P(Outlook = overcast/Play = No)$. This will make the whole probability equivalent to zero. In order to handle this problem, we will use the Laplace estimator.

The new Prior Probabilities will be as follows:

```
Prior Probability P(Play = Yes) = (9 + 1) / (14 + 2)
= 10/16
Prior Probability P(Play = No) = (5 + 1) / (14 + 2)
= 6/16
```

The following table describes the conditional probabilities after the Laplace correction:

Outlook	
$P(\text{Sunny}\|\,\text{Play} = \text{Yes}) = 3/12$	$P(\text{Sunny}\|\,\text{Play} = \text{No}) = 4/8$
$P(\text{Overcast}\|\,\text{Play} = \text{Yes}) = 5/12$	$P(\text{Overcast}\|\,\text{Play} = \text{No}) = 1/8$
$P(\text{Rain}\|\,\text{Play} = \text{Yes}) = 4/12$	$P(\text{Rain}\|\,\text{Play} = \text{No}) = 3/8$
Temperature	
$P(\text{Hot}\|\,\text{Play} = \text{Yes}) = 3/12$	$P(\text{Hot}\|\,\text{Play} = \text{No}) = 3/8$
$P(\text{Mild}\|\,\text{Play} = \text{Yes}) = 5/12$	$P(\text{Mild}\|\,\text{Play} = \text{No}) = 3/8$
$P(\text{Cool}\|\,\text{Play} = \text{Yes}) = 4/12$	$P(\text{Cool}\|\,\text{Play} = \text{No}) = 2/8$
Humidity	
$P(\text{High}\|\,\text{Play} = \text{Yes}) = 4/11$	$P(\text{High}\|\,\text{Play} = \text{No}) = 5/7$
$P(\text{Normal}\|\,\text{Play} = \text{Yes}) = 7/11$	$P(\text{Normal}\|\,\text{Play} = \text{No}) = 2/7$
Windy	
$P(\text{True}\|\,\text{Play} = \text{Yes}) = 4/11$	$P(\text{True}\|\,\text{Play} = \text{No}) = 4/7$
$P(\text{False}\|\,\text{Play} = \text{Yes}) = 7/11$	$P(\text{False}\|\,\text{Play} = \text{No}) = 3/7$

The two probabilities of likelihood can be calculated easily by the following:

P(X/Play = Yes) = P(Outlook = Overcast| Play = Yes) × P(Temperature = Mild| Play = Yes) × P(Humidity = Normal| Play = Yes) × P(Windy = False| Play = Yes)

P(X/Play = Yes) = 5/12 × 5/12 × 7/11 × 7/11 = 0.070305

P(X/Play = No) = P(Outlook = Overcast| Play = No) × P(Temperature = Mild| Play = No) × P(Humidity = Normal| Play = No) × P(Windy= False| Play = No)

P(X/Play = No) = 1/8 × 3/8 × 2/7 × 3/7 = 0.00574

4.6 Posterior Probability

In order to calculate the posterior probability, we need three things.

1. Prior probability
2. Likelihood
3. Evidence

The following formula shows the relationship among the three variables to calculate posterior probability:

$$\text{Posterior} = \frac{\text{Prior} \times \text{Likelihood}}{\text{Evidence}}$$

For the classification purpose, we are interested in calculating and comparing the numerator of the above fraction because the *evidence* in the denominator is same for both classes. In other words, the posterior is proportional to the likelihood times the prior.

$$\text{Posterior} \propto \text{Prior} \times \text{Likelihood}$$

The numerator *prior × likelihood* for two classes can be calculated by simply multiplying the respective prior probabilities and the probabilities of likelihood.

$$P(\text{Play} = \text{Yes}/X) \propto \qquad P(\text{Play} = \text{Yes}) \times P(X/\text{Play} = \text{Yes})$$

$$10/16 \times 0.070305 = 0.043941$$

$$P(\text{Play} = \text{No}/X) \propto \qquad P(\text{Play} = \text{No}) \times P(X/\text{Play} = \text{No})$$

$$6/16 \times 0.00574 = 0.002152$$

Since the value of P(Play = Yes) × P(X/Play = Yes) > P(Play = No) × P(X/Play = No), we will assign class "Yes" to the new case "X."

4.7 MATLAB Implementation

In MATLAB, one can perform calculations related to the naïve Bayesian classifier easily.

We will first load the same dataset that we have discussed in the chapter as an example in the MATLAB environment and we will then calculate different parameters related to the naïve Bayesian classifier.

The following code snippet loads the data from "data.csv" into the MATLAB environment.

```
fid = fopen('C:\Naive Bayesian\data.csv')';
out = textscan(fid,'%s%s%s%s%s%s','delimiter',',');
fclose(fid);
num_featureswithclass = size(out,2);
tot_rec = size(out{size(out,2)},1)-1;
for i = 1:tot_rec
    yy{i} = out{num_featureswithclass}{i+1};
end
for i = 1: num_featureswithclass
    xx{i} = out{i};
end
```

For calculating the prior probabilities of the class variable, the following code snippet will perform the job.

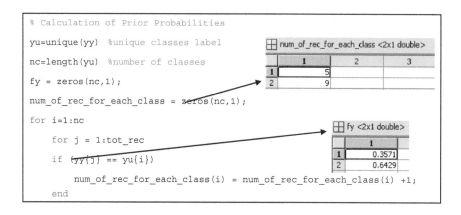

In order to calculate the likelihood table, the following code snippet works:

```
prob_table=zeros(num_featureswithclass-1,10,nc);
for col = 1:num_featureswithclass-1
unique_value = unique(xx{col});
rec_unique_value{col} = unique_value;
    for i = 2:length(unique_value)
        for j = 2:tot_rec+1
        if strcmp(xx{col}{j}, unique_value{i}) == 1 &&
strcmp(xx{num_featureswithclass}{j}, yu{1}) ==1
        prob_table(col, i-1,1) = prob_table(col,
i-1,1) + 1;
        end
        if strcmp(xx{col}{j}, unique_value{i}) == 1 &&
strcmp(xx{num_featureswithclass}{j}, yu{2}) ==1
        prob_table(col, i-1,2) = prob_table(col,
i-1,2) + 1;
        end
        end
    end
end
prob_table(:,:,1) = prob_table(:,:,1)./
num_of_rec_for_each_class(1);
prob_table(:,:,2) = prob_table(:,:,2)./
num_of_rec_for_each_class(2);
```

The matrix "prob_table" used in the above code is a matrix of 4 × 10 × 2 dimension where "4" is the number of attributes in the dataset. The number "10" is the possible number of unique value in any attribute. In this example, the maximum number was "3." The number "2" refers to the number of classes. If we see the values present in the prob_table, the understanding will be further enhanced.

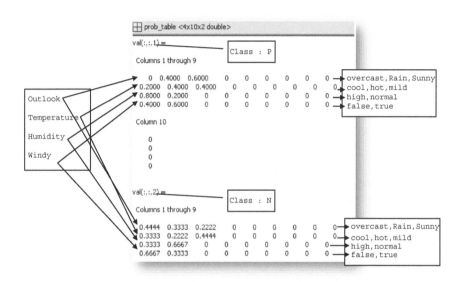

Predicting for an unlabeled record:

Now that we have a naïve Bayesian classifier in the form of tables, we can use them to predict newly arriving unlabeled records. The following code snippet describes the prediction process in MATLAB.

```
A = {'sunny', 'hot','high','false'};
A1 = find(ismember(rec_unique_value{1},A{1}));
  A11 = 1;
A2 = find(ismember(rec_unique_value{2},A{2}));
  A21 = 2;
A3 = find(ismember(rec_unique_value{3},A{3}));
  A31 = 3;
A4 = find(ismember(rec_unique_value{4},A{4}));
  A41 = 4;
```

```
ProbN = prob_table(A11,A1 - 1,1)*prob_
  table(A21,A2 - 1,1) *prob_table(A31,A3 - 1,1)
  *prob_table(A41,A4 - 1,1)*fy(1);
ProbP = prob_table(A11,A1 - 1,2)*prob_
table(A21,A2 - 1,2) *prob_table(A31,A3 - 1,2)
*prob_table(A41,A4 - 1,2) *fy(2);
if ProbN > ProbP
prediction = 'N'
else
prediction = 'P'
end
```

References

1. Good, I. J. *The Estimation of Probabilities: An Essay on Modern Bayesian Methods.* Cambridge: MIT Press, 1965.
2. Kendall, M. G. and Stuart, A. *The Advanced Theory of Statistics.* London: Griffin, 1968.

Chapter 5

The *k*-Nearest Neighbors Classifiers

5.1 Introduction

In pattern recognition, the *k*-nearest neighbors algorithm (or *k*-NN for short) is a nonparametric method used for classification and regression [1]. In both cases, the input consists of the *k* closest training examples in the feature space. The output depends on whether *k*-NN is used for classification or regression:

- In *k*-NN classification, the output is a class membership. An object is classified by a majority vote of its neighbors, with the object being assigned to the class most common among its *k*-NN (*k* is a positive integer, typically small). If $k = 1$, then the object is simply assigned to the class of that single nearest neighbor.
- In *k*-NN regression, the output is the property value for the object. This value is the average of the values of its *k*-NN.

k-NN is a type of instance-based learning, or lazy learning, where the function is only approximated locally and all computation is deferred until classification. The k-NN algorithm is among the simplest of all machine learning algorithms.

For both classification and regression, it can be useful to assign weight to the contributions of the neighbors, so that the nearer neighbors contribute more to the average than the more distant ones. For example, a common weighing scheme consists of giving each neighbor a weight of $1/d$, where d is the distance to the neighbor.

The neighbors are taken from a set of objects for which the class (for k-NN classification) or the object property value (for k-NN regression) is known. This can be thought of as the training set for the algorithm, though no explicit training step is required.

A shortcoming of the k-NN algorithm is that it is sensitive to the local structure of the data. The algorithm has nothing to do with and is not to be confused with k-means, another popular machine learning technique.

5.2 Example

Suppose that we have a two-dimensional data, consisting of circles, squares, and diamonds as in Figure 5.1.

Each of the diamonds is desired to be classified as either a circle or a square. Then, the k-NN can be a good choice to do the classification task.

The k-NN method is an instant-based learning method that can be used for both classification and regression.

Suppose that we are given a set of data points $\{(x_1, C_1), (x_2, C_2), \ldots, (x_N, C_N)\}$, where each of the points x_j, $j = 1, \ldots, N$ has m attributes $a_{j1}, a_{j2}, \ldots, a_{jm}$ and $C_1, \ldots, C_N$ are taken from some discrete or continuous space K.

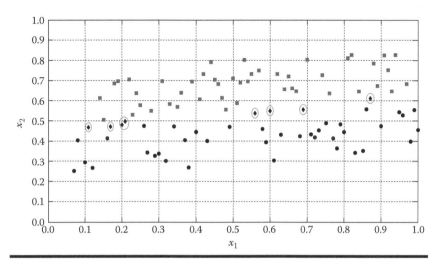

Figure 5.1   Two classes of squares and circle data, with unclassified diamonds.

It is clear that

$$X \xrightarrow{\ f\ } K$$

with $f(x_j) = C_j$, where $X = \{x_1, \ldots, x_N\}$ is a subset of some space Y.

Given an unclassified point $x_s = (a_{s1}, a_{s2}, \ldots, a_{sm}) \in Y$, we would like to find $C_s \in K$ such that $f(x_s) = C_s$. At this point, we have two scenarios [2]:

1. *The space K is discrete and finite*: in this case we have a classification problem, $C_s \in K$ where $K = \{C_1, \ldots, C_N\}$, where and the k-NN method sets $f(x_s)$ to be the major vote of the k-NN of x_s.
2. *The space K is continuous*: In this case, we have a regression problem and the k-NN method sets $f(x_s) = (1/k)\sum_{j=1}^{k} f(x_{s_j})$, where $\{x_{s1}, x_{s2}, \ldots, x_{sk}\}$ is the set of k-NN of x_s. That is, $f(x_s)$ is the average of the values of the k-NN of point x_s.

The k-NN method belongs to the class of instance-based supervised learning methods. This class of methods does not create an approximating model from a training set as happens in the model-based supervised learning methods.

To determine the k-NN of a point x_s, a distant measure must be used to determine the k closest points to point x_s: $\{x_{s_1}, x_{s_2}, \ldots, x_{s_k}\}$. Assuming that $d(x_s, x_j)$, $j = 1, \ldots, N$, measures the distance between x_s and x_j, and $\{x_{s_i} : i = 1, \ldots, k\}$ is the set of k-NN of x_s according to the distant metric d. Then, the approximation $f(x_s) = (1/k)\sum_{j=1}^{k} f(x_{s_j})$ assumes that all the k-neighboring points have the same contribution to the classification of the point x_s.

5.3 *k*-Nearest Neighbors in MATLAB®

MATLAB enables the construction of a k-NN method through the method "ClassifyKNN.fit," which receives a matrix of attributes and a vector of corresponding classes. The output of the ClassifyKNN.fit is a k-NN model. The default number of neighbors is 1, but it is possible to change this number through setting the attribute "NumNeighbors" to the desired value.

The following MATLAB script applies the k-NN classifier to the ecoli dataset:

```
clear; clc;
EcoliData = load('ecoliData.txt'); % Loading the
ecoli dataset
EColiAttrib = EcoliData(:, 1:end-1); % ecoli
attributes
EColiClass = EcoliData(:, end); % ecoli classes
%knnmodel = ClassificationKNN.
fit(EColiAttrib(1:280,:), EColiClass(1:280),...
% 'NumNeighbors', 5, 'DistanceWeight', 'Inverse');
% fitting the
% ecoli data with the k-nearest neighbors method
% The above line changes the number of neighbors
to 4
```

```
knnmodel = ClassificationKNN.
fit(EColiAttrib(1:280,:), EColiClass(1:280),...
'NumNeighbors', 5)
Pred = knnmodel.predict(EColiAttrib(281:end,:));
knnAccuracy = 1-find(length(EColiClass(281:end)-
Pred))/length(EColiClass(281:end));
knnAccuracy = knnAccuracy * 100
```

The results are as follows:

```
knnmodel =
  ClassificationKNN
    PredictorNames: {'x1' 'x2' 'x3' 'x4' 'x5'
    'x6' 'x7'}
    ResponseName: 'Y'
     ClassNames: [1 2 3 4 5 6 7 8]
    ScoreTransform: 'none'
    NObservations: 280
        Distance: 'euclidean'
    NumNeighbors: 5
   Properties, Methods
  knnAccuracy =
```

`98.2143`

Another approach assumes that a closer point to x_s shall have more contribution to the classification of x_s. Therefore, came the idea of the weighted weights, which is assumed to be proportional to the closeness of the point from x_s. Now, if $\{x_{s_1}, x_{s_2}, \ldots, x_{s_k}\}$ denotes the k-NN of x_s, let

$$w(x_s, x_{s_j}) = \frac{e^{-d(x_s, x_{s_j})}}{\sum_{i=1}^{k} e^{-d(x_s, x_{s_i})}}$$

It is obvious that $\sum_{i=1}^{k} w(x_s, x_{s_i}) = 1$. Finally, $f(x_s)$ is approximated as

$$f(x_s) = \frac{1}{k} \sum_{j=1}^{k} w(x_s, x_{s_j}) \cdot f(x_{s_j})$$

MATLAB enables the use of weighted distance through changing the attribute "DistanceWeights" from "equal" to either "Inverse" or "SquaredInverse." This is done by using

```
knnmodel = ClassificationKNN.
fit(EColiAttrib(1:280,:), EColiClass(1:280),...
'NumNeighbors', 5, 'DistanceWeight', 'Inverse');
```

Applying the KNN classifier with a weighted distance provides the following results:

```
knnAccuracy =
98.2143
```

which is the same as the model with equal distance.

References

1. Hastie, T., Tibshirani, R., and Friedman, J. *The Elements of Statistical Learning: Data Mining, Inference, and Prediction*, 2nd Ed., New York: Springer-Verlag, February 2009.
2. Vapnik, V. N. *The Nature of Statistical Learning Theory.* 2nd Ed., New York: Springer-Verlag, 1999.

Chapter 6

Neural Networks

6.1 Perceptron Neural Network

A neural network is a model of reasoning that is inspired by biological neural networks, which is the central nervous system in an animal brain. The human brain consists of a huge number of interconnected nerve cells called *neurons*.

A neuron consists of a cell body, soma, a number of fibers called *dendrites*, and a single long fiber called the *axon* [1] (Figure 6.1).

The main function of dendrites is to receive messages from other neurons. Then, the signal travels to the main cell body, known as the *soma*. The signal leaves the soma and travels down the axon to the synapse. The message then moves through the axon to the other end of the neuron, then to the tips of the axon and then into the space between neurons. From there, the message can move to the next neuron.

The human brain incorporates nearly 10 billion neurons and 60 trillion connections, synapses, between them. By using a massive number of neurons simultaneously, the brain can process data and perform its function very fast [2].

The structure of the biological neural system, together with how it performs its functions has inspired the idea of

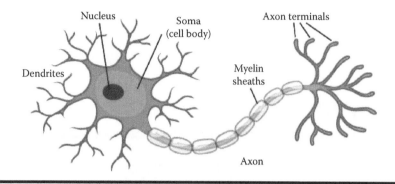

Figure 6.1 A biological neuron.

artificial neural networks (ANNs). The first neural network conceptual model was introduced in 1943 by Warren McCulloch and Walter Pitts. They described the concept of neuron as an individual cell that communicates with other cells in a network. This cell receives data from other cells, processes the inputs, and passes the outputs to other cells. Since then, scientists and researchers have made intensive research to develop the ANNs. Nowadays, ANNs are considered one of the most efficient pattern recognition, regression, and classification tools [3].

The big developments in ANNs during the past few decades have motivated human ambitions to create intelligent machines with human-like brain. Many Hollywood movies are based on the idea that human-like smart machines will aim at controlling the universe (*Artificial Intelligence*; *The Matrix*; *The Terminator*; *I, Robot*; *Star Wars*; *Autómata*; etc.). The robots of *I, Robot* are designed to have artificial brains consisting of ANNs.

However, despite the fact that performance of the ANNs is still so far from the human brain, to date the ANNs are one of the leading computational intelligence tools.

6.1.1 Perceptrons

A perceptron is the simplest kind of ANNs, invented in 1957 by Frank Rosenblatt. A perceptron is a neural network

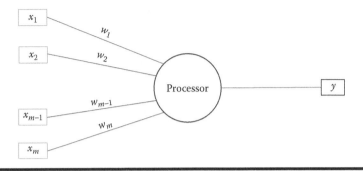

Figure 6.2   A perceptron with *m* input features $(a_1,...,a_m)$.

that consists of a single neuron that can receive multiple inputs and produces a single output (Figure 6.2).

Perceptrons are used to classify linearly separable classes, through finding any *m*-dimensional hyperplane in the feature space that separates the instances of the two classes. In a perceptron model, the weighted sum

$$\sum_{j=1}^{m} w_j \cdot x_j = w_1 \cdot x_1 + \cdots + w_m \cdot x_m$$

is evaluated and passed to an activation function, which compares it to a predetermined threshold θ. If the weighted sum is greater than the threshold θ, then the perceptron fires and outputs 1, otherwise it outputs 0. There are many kinds of activation functions that can be used with the perceptron, but the step, sign, linear, and sigmoid functions are the most popular ones. The step function is of the following form:

$$f(x) = \begin{cases} 1 & \text{if } x > 0 \\ 0 & \text{if } x < 0 \end{cases}$$

The sign function is given by

$$f(x) = \begin{cases} 1 & \text{if } x > 0 \\ -1 & \text{if } x < 0 \end{cases}$$

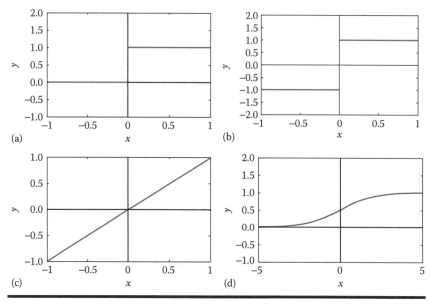

Figure 6.3 Activation functions: (a) step activation function, (b) sign activation function, (c) linear activation function, and (d) sigmoid activation function.

The linear function is $f(x) = x$ and the sigmoid function is (Figure 6.3):

$$f(x) = \frac{1}{1 + e^{-x}}$$

All the above-mentioned activation functions are triggered at a threshold $\theta = 0$. However, it is more convenient to have a threshold other than zero. For doing that, a bias b is introduced to the perceptron in addition to the m inputs $x_1, x_2, \ldots, x_n$. The perceptron can have an additional input called the *bias*. The role of the bias b is to move the threshold function to the left or right, in order to change the activation threshold (Figure 6.4).

Changing the value of the bias b does not change the shape of the activation function, but together with the other weights, it determines when the perceptron fires. It is worthy to note that the input associated with the bias is always one.

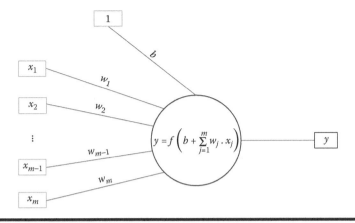

Figure 6.4   A perceptron with *m* inputs and *a* bias.

Now, training the perceptron aims at determining the optimal weights and bias value at which the perceptron fires.

In the given classified data $(x_j, y_j), j = 1,\ldots,N$, each feature vector x_j has m features $(a_{j1},\ldots,a_{jm})$. Each feature vector belongs to either a class C_1 or a class C_2 and $y_j \in \{-1,1\}$. If the two given classes are linearly separable, then the perceptron can be used to classify them. We will assume that instances that belong to class C_1 are classified as 1, whereas instances that belong to class C_2 are classified as −1. Therefore, we will consider the sign activation function for the perceptron to fire. The MATLAB®'s function sign (x) returns 1 if x is positive and −1 if x is negative.

Given a training feature vector $x_j = (a_{j1},\ldots,a_{jm})$ classified as y_j, we are going to show how the perceptron will deal with the vector x_j during the training stage. The first thing we will do is to set

$$
w = \begin{bmatrix} w_1 \\ \vdots \\ w_m \\ b \end{bmatrix}, \quad x_j = \begin{bmatrix} a_{j1} \\ \vdots \\ a_{jm} \\ 1 \end{bmatrix}
$$

and generate random values for the $(m + 1)$-dimensional vector w. We notice that

$$\boldsymbol{w}^T \cdot \boldsymbol{x}_j = w_1 a_{j1} + \cdots + w_m a_{jm} + b$$

is the weighted sum. Since we are interested in the sign of the weighted sum, we can use the MATLAB's sign function to determine the sign of $\boldsymbol{w}^T \cdot \boldsymbol{x}_j$. Now, we define the error in classifying $\boldsymbol{x}_j$ to be the value:

$$E_j = \boldsymbol{y}_j - \mathrm{sign}\left(\boldsymbol{w}^T \cdot \boldsymbol{x}_j\right)$$

The error E_j could be either 2,0 or −2. Both the first and third values of the error (nonzero values) indicate the occurrence of an error in classifying the feature vector $\boldsymbol{x}_j$. Knowing the error helps us in the adjustment of weights.

To readjust the weight, we define a learning rate parameter $\alpha \in (0,1)$. This parameter determines how fast the weights are changed, and hence, how fast the perceptron learns during the training phase. Given the learning rate α, the correction in the weight vector is given by

$$\Delta \boldsymbol{w} = \alpha \boldsymbol{x}_j E_j$$

and the new weight becomes

$$\boldsymbol{w}_{\mathrm{new}} = \boldsymbol{w}_{\mathrm{old}} + \Delta \boldsymbol{w}$$

Applying the round of constructing the weighted sum, evaluating the error, and adjusting the weight vector to all the instances in the training set is called an *epoch*. The perceptron learning process shall consist of an optimal number of epochs.

6.2 MATLAB Implementation of the Perceptron Training and Testing Algorithms

The following MATLAB function applies the perceptron learning:

```
function w = Perceptron Learning(TrainingSet,
Class, Epochs, LearningRate)
%% The output is an (m+1)-dimensional vector of
weights
%% TrainingSet is an n by m matrix, where the rows
of the TrainingSet matrix
%% represent the instances, the columns represent
the features
%% Class is an n-dimensional vector of 1's and
-1's, corresponding to the
%% instances of the training set. Epochs determine
the number of epochs and
%% Learning rate determine the rate at which the
weights are corrected.
[n, m] = size(TrainingSet);
w = 0.5*rand(1, m); % initializing the weights
a = LearningRate;
for epoch = 1: Epochs
  for j = 1: n
    x = TrainingSet(j,:); % Picking an instance x_j
from the training
                    % set
    wsum = sum(w.*x); % Constructing the weighted sum
    if wsum > 0
       y = 1;
    else
       y = -1;
    end
    Error = Class(j) - y; % Error is the difference
    between the
                  % predicted class and actual class
    w = w + Error*x*a; % Correcting the weights
    according the the
                    % error
  end
end
```

For the testing of the algorithm, the PerceptronTesting function can be used. Following is the code for the PerceptronTesting function:

```
function [PredictedClass, Accuracy] =
PerceptronTesting(TestingSet, Class, w)
%% The outputs are a vector of predicted classes
and the prediction
%% accuracy as a percentage. The Accuracy is the
percentage of the ratio
%% between the correctly classified instances in
the testing set and the
%% total number of instances in the testing set.
%% TestingSet is an N by m matrix and Class are the
corresponding classes
%% for the feature vectors in the testing set
matrix. The vector w is the
%% the vectors of weights, obtained during the
training phase
[N, m] = size(TestingSet);
PredictedClass = zeros(N, 1);
for j = 1: N
  x = TestingSet(j,:);
  wsum = sum(w.*x);
  if wsum > 0
     PredictedClass(j) = 1;
  else
     PredictedClass(j) = -1;
  end
end
Error = Class - PredictedClass;
Accuracy = (1 - length(find(Error))/
length(Error))*100;
```

6.3 Multilayer Perceptron Networks

A single perceptron can solve any classification problem for linearly separable classes. If given two nonlinearly separable classes, a single layer perceptron network will fail to solve the problem of classifying them. The most common simple non-linearly separable problem is the logical XOR problem. The XOR logical operation takes two logical inputs x and y, then the output of the XOR operator is as given in the following table (Table 6.1).

The output of the XOR operation is either 0 or 1, but the two classes are not separable in the feature space (Figure 6.5).

From the Figure 6.5, it is obvious that the true instances cannot be separated from the false instances by using a straight line. Such a nonlinearly separable problem is solved by using a multilayer perceptron network. Indeed, the decision boundaries

Table 6.1   XOR Logical Operation

x	y	$x \oplus y$
0	0	1
0	1	0
1	0	0
1	1	1

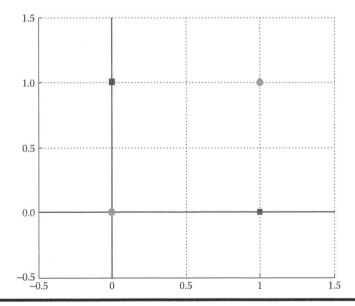

Figure 6.5   The XOR binary operation: True instances are plotted in circles and false instances are plotted with squares.

in a multilayer perceptron network have a more complex geo-metric shape in the feature space than in a hyperplane.

In a multilayer perceptron neural network, each perceptron receives a set of inputs from other perceptrons, and according to whether the weighted sum of the inputs is above some threshold value, it either fires or does not. As in a single perceptron network, the bias (which determines the threshold) in addition to the weights are adjusted during the training phase (Figure 6.6).

At the final neural network model, for some input, a spe-cific set of the neurons fire. Changing the input changes the set of neurons that fires. The main purpose from the neural network training is to learn when to fire each neuron as a response to a specific input.

To learn a neural network, random weights and biases are generated. Then, a training instance is passed to the neural network, where the output of each layer is passed to the next layer until computing the predicted output at the output layer, according to the initial weights. The error at the output layer is computed as the difference between the actual and pre-dicted outputs. According to the error, the weights between the output layer and the hidden layers are corrected, and then

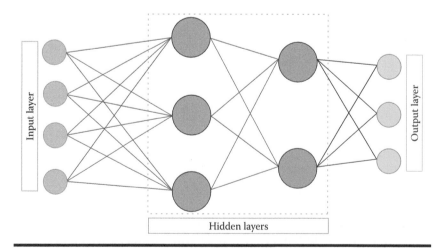

Figure 6.6　A multilayer perceptron neural network.

the weights between the hidden layer and the input layer are adjusted in a backward fashion. Another training instance is passed to the neural network and to the process of evaluating the error at the output layer, thereby correcting the weights between the different layers from the output layer to the input layer. Repeating this process for as many epochs will help in learning the neural network.

6.4 The Backpropagation Algorithm

The backpropagation algorithm consists of two stages:

Step 1: This step is called the *feed-forward stage*; and at this stage the inputs are fed to the network and the output is computed at both the hidden and output layers.

Step 2: The prediction error is computed at the output layer, and this error is propagated backward to adjust the weights. This step is called the *backpropagation*. The backpropagation algorithm uses deterministic optimization to minimize the squared error sum using the gradient decent method. The gradient decent method requires computing the partial derivatives of the activation function with respect to the weights of the inputs. Therefore, it is not applicable to use the hard limit activation functions (step and sign functions).

Generally, a function $f: R \rightarrow [0,1]$ is an activation function if it satisfies the following properties:

1. The function f has a first derivative f'.
2. The function f is a nondecreasing function, that is, $f'(x) > 0$ for all $x \in R$.
3. The function f has horizontal asymptotes at both 0 and 1.
4. Both f and f' are computable functions.

The sigmoid functions are the most popular functions in the neural networks. Two sigmoid functions are usually used as activation functions:

1. $S_1(x) = 1/(1 + e^{-x})$

2. $S_2(x) = (1 - e^{-x})/(1 + e^{-x})$

The derivatives of the two functions $S_1(x)$ and $S_2(x)$ are given by

$$\frac{dS_1(x)}{dx} = S_1(x)(1 - S_1(x))$$

and

$$\frac{dS_2(x)}{dx} = 1 - S_2(x)$$

The second sigmoid function $S_2(x)$ is the hyperbolic tangent function. Choosing sigmoid functions guarantees the continuity and differentiability of the error function. The graphs of the two sigmoid functions $S_1(x)$ and $S_2(x)$ are explained in Figure 6.7.

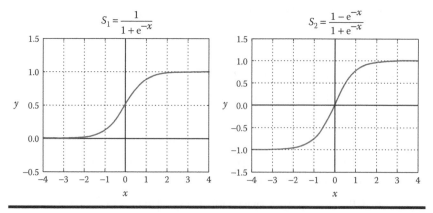

Figure 6.7   The two sigmoid functions $S_1(x)$ and $S_2(x)$.

6.4.1 Weights Updates in Neural Networks

Given is the classified data $(x_j, y_j), j = 1,\ldots,N$, where each feature vector x_j has m features $(a_{j1},\ldots,a_{jm})$. Let p_j be the predicted output by the neural network, when presenting a feature vector x_j. Then, the error is given as follows:

$$E = \frac{1}{2} \sum_{j=1}^{N} \left\| y_j - p_j \right\|^2$$

The backpropagation algorithm works to find a local minimum of the error function, where the optimization process runs over the weights and biases of the neural network. It is noteworthy to remember that the biases are embedded in weights. If the neural network consists of a total of L weights, then the gradient of the error function with respect to the network weights is

$$\nabla E = \begin{bmatrix} \dfrac{\partial E}{\partial w_1} \\ \vdots \\ \dfrac{\partial E}{\partial w_L} \end{bmatrix}$$

In the gradient decent method, the update in the weights vector is proportional to negative of the gradient. That is,

$$\Delta w_j = -\alpha \frac{\partial E}{\partial w_j}, \quad j = 1,\ldots,L$$

where, $\alpha > 0$ is a constant, representing the learning rate. We will assume that the activation function $S_1(x)$ is used throughout the network. Then, at a unit i, that receives inputs $(z_1,\ldots,z_M)$ with weights $(w_{i1},\ldots,w_{iM})$ gives an output o_i, where:

$$o_i = S_1\left(\sum_{k=1}^{M} w_{ik} z_k \right)$$

If the target output of unit i is u_i, then by using the chain rule, we get:

$$\frac{\partial E}{\partial \pmb{w}_{ij}} = \frac{\partial E}{\partial o_i} \cdot \frac{\partial o_i}{\partial \pmb{w}_{ij}} = -\left(u_i - o_i\right)o_i\left(1 - o_i\right)z_j$$

Therefore, the correction on the weight w_{ij} is given as follows:

$$\Delta \pmb{w}_{ij} = \alpha\left(u_i - o_i\right)o_i\left(1 - o_i\right)z_j$$

The weights are first adjusted at the output layer (weights from the hidden layer to the output layer) and then adjusted at the hidden layer (from the input layer to the hidden layer), assuming that the network has one hidden layer.

6.5 Neural Networks in MATLAB

MATLAB enables the implementation of a neural network through its neural network toolbox [4]. The initialization of the neural network is done through the MATLAB command feedforwardnet. The following MATLAB script is used to train a network with the "ecoli" dataset (Figure 6.8).

```
clear; clc;
A = load('EColi1.txt'); % Loading the ecoli data,
with the classes at the
% last column
%B = A(1:end, 2:end);
C = A(1:end, 1:end-1)'; % C is the matrix of the
feature vectors
T = A(:, end)'; % T is the vector of classes
net = feedforwardnet; % Initializing a neural
network 'net'
net = configure(net, C, T);
hiddenLayerSize = 10; % Setting the number of
hidden layers to 10
```

```
net = patternnet(hiddenLayerSize); % Pattern
recognition network
net.divideParam.trainRatio = 0.7; % Ratio of
training data is 70%
net.divideParam.valRatio = 0.2; % Ratio of
validation data is 20%
net.divideParam.testRatio = 0.1; % Ratio of testing
data is 10%
[net, tr] = train(net, C, T); % Training the
network and the resulting
                       % model is the output net
outputs = net(C); % applying the model to the data
errors = gsubtract(T, outputs); % computing the
classification errors
performance = perform(net, T, outputs)
view(net)
```

The outputs of the above script are as follows:
```
>> performance =
  0.7619
```

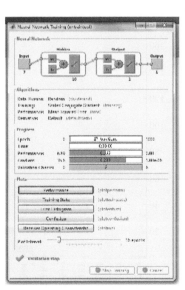

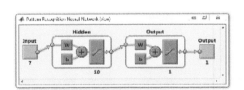

**Figure 6.8   The output neural network model for classifying the
ecoli data.**

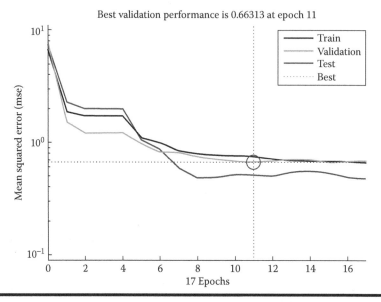

Figure 6.9   The validation performance of the neural network model.

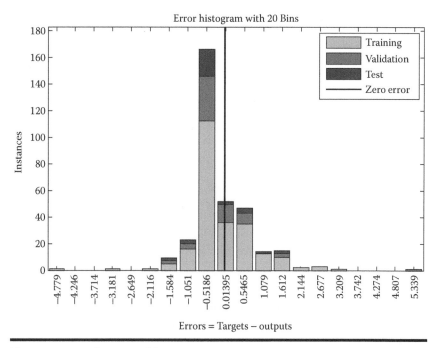

Figure 6.10   The error histogram.

A full report about the trained model can be obtained by the neural network toolbox. The information included in the report includes the algorithms, progress, and plots. The algorithms indicate the method by which the data are divided, the optimization algorithm used for the minimization of the error, and the type of the error. The progress tells the number of epochs to reach the optimal solution, the time consumed, the performance, gradient, and validation checks. The plots include five possible plots, including the performance, the training state through the different epochs, the errors, the confusion matrices, and the receiver operating characteristics (ROCs). The performance of the neural network can be viewed through the performance button (Figure 6.9).

The error histogram can also be seen through the error histogram button (Figure 6.10).

References

1. Vapnik, V. N. *The Nature of Statistical Learning Theory*, 2nd Ed., New York: Springer-Verlag, 1999.
2. Negnevitsky, M. *Artificial Intelligence: A Guide to Intelligent Systems*, 2nd Ed., London: Pearson Education Limited, 2005.
3. Manning, C. D., Raghavan, P., and Schutze, H. *Introduction to Information Retrieval.* New York: Cambridge University Press, 2008.
4. Beale, M. H., Hagan, M. T., and Demuth, H. B. *Neural Network Toolbox User's Guide.* Natick, MA: The MathWorks, 2015.

Chapter 7

Linear Discriminant Analysis

7.1 Introduction

In 1936, statistical pioneer Ronald Fisher discussed linear discriminant [1] that became a common method to be used in statistics, pattern recognition, and machine learning. The idea was to find a linear combination of features that are able to separate two or more classes. The resulting linear combination can also be used for dimensionality reduction. Linear discriminant analysis (LDA) is a generalization of the Fisher linear discriminant.

This method was used to explain the bankruptcy or survival of the firm [2]. In face recognition problems, it is used to reduce dimensions.

LDA seeks to maximize class discrimination and produces exactly as many linear functions as there are classes. The predicted class for an instance will be the one that has the highest value for its linear function.

7.2 Example

Let us say we want to predict the type of smartphone a customer would be interested in. Different smartphones will be the classes and the known data related to customers will be represented by x.

In order to concoct two class problems, we will define two classes as "Apple" and "Samsung."

```
C = {"Apple", "Samsung"}
```

We will represent Apple as 0 and Samsung as 1 for the type of smartphones in our practical implementation, respectively, or $C = \{0,1\}$.

The two numeric variables that will be considered to predict the classes are *age* and *income* of customers. The variable x_1 will represent the age of the customer and x_2 will represent the income of the customer.

μ_i is the vector that will describe the mean age and mean income of the customers of smartphones of type i. Σ_i *will* be the covariance matrix of age and income for type i.

We will randomly generate the data for 25 customers in order to understand how the classification works using discriminant analysis.

```
X1_Apple_Age = round(30 + randn(10,1)*5);
% Supposition that average age of Apple buyers is
30 and a standard deviation of 5
X1_Samsung_Age = round(45 + randn(15,1)* 10);
% Supposition that average age of Samsung buyers is
45 and a standard deviation of 10
X2_Apple_income = round(10000 + randn(10,1) *
2000); % Supposition that average income of Apple
buyers is $10000 and a standard deviation of $2000
X2_Samsung_income = round(5000 + randn(15,1) * 500);
% Supposition that average income of Samsung buyers
is $5000 and a standard deviation of $500
```

```
X1 = [X1_Apple_Age; X1_Samsung_Age];
X2 = [X2_Apple_income; X2_Samsung_income];
X = [X1 X2];
```

To assign the class to the 25 records, we will simply use the following MATLAB® code:

```
Y = [zeros(10,1); ones(15,1)] % Assign first
10 rows the value of 0 (or Apple) and the last
15 rows the value of 1 (represent Samsung)
```

To visualize the above data, the following MATLAB code can be used:

```
scatter(X(1:10,1),X(1:10,2),'r+') % red +
representing data related to first group or Apple
category
hold on;
scatter(X(11:25,1),X(11:25,2),'b^') % blue
^representing data related to second group or
Samsung category
```

The above code will result in following Figure 7.1.

To perform discriminant analysis, we will first initialize few variables that will be used in the discrimination process.

```
[rows columns] = size(X); % Determine number of
rows and columns of input data

Labels = unique(Y); % Label will contain the two
unique values of Y
k = length(Labels); % k will contain number of
records for each label

% Initialize
nClass     = zeros(k,1);   % Class counts
ClassMean = zeros(k, columns);   % Class sample
means
PooledCov = zeros(columns, columns);   % Pooled
covariance
Weights    = zeros(k, columns+1);   % model
coefficients
```

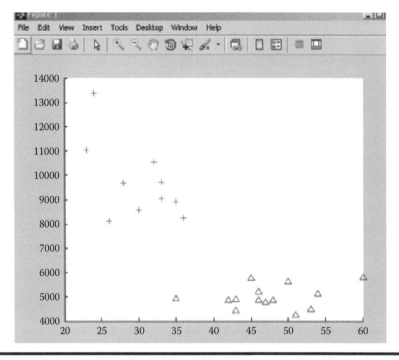

Figure 7.1   Result of MATLAB code represents two classes.

In order to calculate weights that will be used for classification, we will have to calculate the mean vector as well as the covariance matrix. The covariance matrix of the two groups of data belonging to the different classes can be calculated simply by using the "cov()" command of MATLAB.

The following MATLAB code describes mean and covariance matrix calculation of the two groups of data.

```
Group1 = (Y == Labels(1)); % i.e class equal to 0
Group2 = (Y == Labels(2)); % i.e class equal to 1

% Group1 and Group2 are Boolean arrays with 1s and
0s.
% In order to find how many items in each group
are, we
% will convert them to number and then sum them.
```

```
numGroup1= sum(double(Group1));
numGroup2= sum(double(Group2));

MeanGroup(1,:) = mean(X(Group1,:)); %Find mean
vector for class 0
MeanGroup(2,:) = mean(X(Group2,:)); %Find mean
vector for class 1

Cov1 = cov(X(Group1,:)); % Covariance matrix
calculation for class 0
Cov2 = cov(X(Group2,:)); % Covariance matrix
calculation for class 1
```

In order to illustrate the calculation, it is better to show the original data along with the associate mean and the covariance matrix:

Age	Income	Class		Age	Income	Class	
23	11030	0		43	4413	1	
32	10523	0		43	4904	1	
26	8117	0		42	4863	1	
28	9675	0		45	5765	1	
33	9708	0		46	4875	1	
35	8936	0		53	4468	1	
24	13364	0		60	5802	1	
36	8249	0		50	5617	1	
33	9032	0		43	4885	1	
30	8576	0		51	4247	1	
				47	4778	1	
				35	4922	1	
				54	5138	1	
				48	4869	1	
				46	5222	1	

Mean age and income for Class 0 can be calculated easily and is as follows:

30   9721

Similarly, mean age and income for Class 1 is given as follows:

47.066   4984.53

The variable *MeanGroup* will hold these two mean vectors in the form of matrix.

30	9721
47.066	4984.53

The two variables *Cov1* and *Cov2* will hold the data related to covariance matrix of data belonging to the two classes:

20.8888888888889	4196.88888888889
4196.88888888889	2530221.11111111

35.9238095238095	722.104761904762
722.104761904762	214130.266666667

Rather than using two covariance matrices, we will pool the data and estimate a common covariance matrix (a technique discussed in the machine learning literature) for all classes. The following MATLAB code describes the calculation.

```
PooledCov = (numGroup1-1)/(rows-k).
*Cov1+(numGroup2-1)/(rows-k).*Cov2
```

Pooled covariance matrix with 9/23 part of Cov1 and 14/23 part of Cov2:

30.0405797101449	1202.71884057971
1202.71884057971	1120426.68405797

We also have to calculate the prior probabilities of the two groups.

```
PriorProb1 = numGroup1 / rows; % The prior
probability of Class 0
PriorProb2 = numGroup2/ rows; % The prior
probability of Class 1
```

Variable PriorProb1 will be simply calculated as 10/25 or 0.4, whereas the second variable PriorProb2 will have a value of 15/25 or 0.6.

Now with all above calculations, we want to calculate the weights that will be used for classification purpose. The following MATLAB code will calculate the weights for us.

```
Weights(1,1)= -0.5*(MeanGroup(1,:)/PooledCov)*
MeanGroup(1,:)' + log(PriorProb1);

Weights(1,2:end) = MeanGroup(1,:)/PooledCov;

Weights(2,1)= -0.5*(MeanGroup(2,:)/PooledCov)*
MeanGroup(2,:)' + log(PriorProb2);

Weights(2,2:end) = MeanGroup(2,:)/PooledCov;
```

The above code yields the following values for the weight matrix:

W0	W1	W2
−71.5217983501704	1.40645733832605	0.0101859165812994
−59.3830490712585	1.82324057744366	0.00640593376510738

These are the weights that LDA will use for classification.

References

1. Fisher, R. A. The use of multiple measures in taxonomic problems, *Annals of Eugenics*, vol. 7, 179–188, 1936.
2. Altman, E. I. Financial ratios, discriminant analysis and the prediction of corporate bankruptcy, *The Journal of Finance*, vol. 23, issue 3, 589–609, 1968.

Chapter 8

Support Vector Machine

8.1 Introduction

In machine learning, support vector machines (SVMs; also, support vector networks [1]) are supervised learning models with associated learning algorithms that analyze data and recognize patterns used for classification and regression analysis. Given a set of training examples, each marked for belonging to one of two categories, an SVM training algorithm builds a model that assigns new examples into one category or the other, making it a nonprobabilistic binary linear classifier. An SVM model is a representation of the examples as points in space, mapped, so that the examples of the separate categories are divided by a clear gap that is as wide as possible. New examples are then mapped into that same space and predicted to belong to a category based on which side of the gap they fall on [2, p.115].

In addition to performing linear classification, SVMs can efficiently perform a nonlinear classification using what is called the *kernel trick*, implicitly mapping their inputs into high-dimensional feature spaces.

When data is not labeled, a supervised learning is not possible, and an unsupervised learning is required that would find natural clustering of the data to groups, and map new data to these formed groups. The clustering algorithm that provides an improvement to the SVMs is called *support vector clustering*, which is highly used in industrial applications either when data is not labeled or when only some data is labeled as a preprocessing for a classification pass; the clustering method was published.

8.2 Definition of the Problem

Given N vectors $\{x_1, x_2, ..., x_N\}$, with each vector x_j has m features $\{a_{j1}, a_{j2}, ..., a_{jm}\}$ and belongs to one of two classes C_1 or C_2. We say that the given data is *linearly separable* if it is possible to find a hyperplane in the feature space that can separate between the instances from class C_1 and the instances from class C_2 [3, p.116].

In Figure 8.1, we show linearly separable data (circles and squares) in a two-dimensional feature space. The discrete line represents the separating hyperplane, which is a straight line in two dimensions.

In Figure 8.2, we show a three-dimensional (3D) hyperplane, separating between the squares and the circles classes. The hyperplane in 3D is a plane in the feature space, represented by the features x_1, x_2, and x_3. Data in spaces with dimensions bigger than three cannot be visualized, but the same concept of separating between features from two different classes with a hyperplane can be applied.

If the given data is not linearly separable in the m-dimensional feature space, it is possible to map it into a higher dimensional space, wherein the given data becomes linearly separable. Therefore, throughout the chapter, we will assume that the given data is linearly separable [4, p.116].

Because of the linearly separable data, there will be a gap between the instances of the two given classes. This gap is referred to as the *margin*. It would be an advantage if the

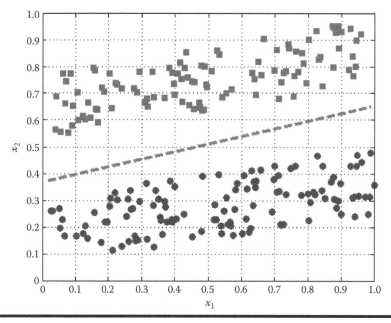

Figure 8.1 Linearly separable data of circles and squares in a two-dimensional feature space.

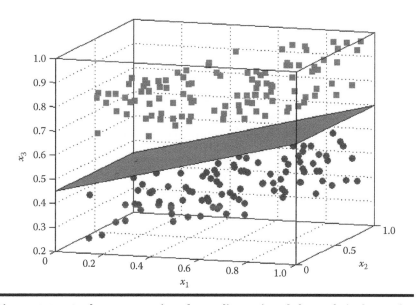

Figure 8.2 A plane separating three-dimensional data of circles and squares.

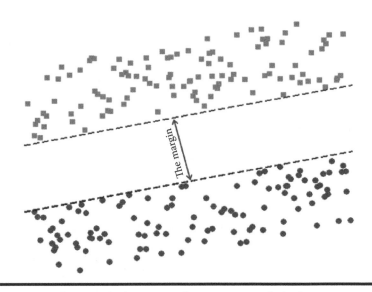

Figure 8.3   The margin between the squares and circles.

margin between the instances of the two classes is as wide as possible (Figure 8.3).

Suppose that we randomly select *n* feature vectors of the given *N* feature vectors, to be a training set. The main goal of the SVM is to find a hyperplane that classifies all training vectors in the two classes. There could be many hyperplanes that separate the instances of the two classes, but it is desired that the chosen hyperplane maximizes the margin between the two classes.

The SVM, $S(x)$, is a linearly discriminant function of the form:

$$S(x) = w^T x + b$$

where x is a feature vector, w is an *m*-dimensional weighting vector, and the scalar b is a bias. The weight vector is orthogonal to the hyperplane and controls its direction, whereas the bias controls its position.

Now, if given a feature vector $\tilde{x}$, then $S(\tilde{x}) = w^T \tilde{x} + b$ to satisfy:

1. $S(\tilde{x}) = w^T\tilde{x} + b > 0$ if $\tilde{x}$ is an instance in C_1.
2. $S(\tilde{x}) = w^T\tilde{x} + b < 0$ if $\tilde{x}$ is an instance in C_2.

The training stage of the SVM includes adjusting the weight and the bias, such that all the instances of C_1 lie on one side of the hyperplane, and the instances of C_2 lie on the other side of the hyperplane.

It is noteworthy that if the position of the hyperplane is set to be close to the instances of C_1, for example, then we subject an instance of C_1 to a small perturbation that can cause it to move to the other side of the hyperplane and hence, being classified wrongly as a C_2 instance. The same applies if the position of the hyperplane is set to be close to instances from C_1. Therefore, the optimal choice for the position of the hyperplane is to be as far as possible from the closest instances from both classes C_1 and C_2. Those closest feature vectors to the hyperplane from both the classes are referred to as the *support vectors* (Figure 8.4).

The main objective of the SVM is to maximize the margin, which is the distance between the hyperplane and the closest vectors to it from both classes (support vectors), such that instances from classes C_1 and C_2 are equally apart from the hyperplane.

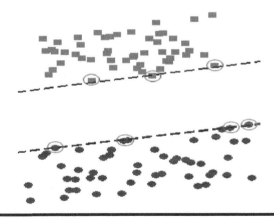

Figure 8.4 Support vectors from the two classes are surrounded by circles.

8.2.1 Design of the SVM

Given a feature vector x_i, then $w^T x_i + b$ is positive if $x_i \in C_1$ and is negative if $x_i \in C_2$. Now, for each x_i we choose y_i, such that

$$y_i = \begin{cases} 1 & \text{if } x_i \in C_1 \\ 0 & \text{if } x_i \in C_2 \end{cases}$$

Under this choice for y_i, we find that

$$y_i \left(w^T x_i + b \right) > 0$$

If x is any point in the feature space, then its distance from the hyperplane is given by

$$z = \frac{\left| w^T x + b \right|}{\| w \|}$$

If the distance between the point x and the hyperplane is desired to be greater than some positive value δ, then we will have the following inequality:

$$\frac{\left| w^T x + b \right|}{\| w \|} \geq \delta$$

From which we get the inequality: $\left| w^T x + b \right| > \delta \| w \|$. It is possible to choose the parameter δ such that $\delta \| w \| = 1$ and we obtain the following inequality:

$$\left| w^T x + b \right| \geq 1$$

Follows from the above inequality that

1. $w^T x + b \geq 1$ if $x \in C_1$
2. $w^T x + b \leq -1$ if $x \in C_2$

The closest feature vectors to the hyperplane from both classes C_1 and C_2 are referred to as *support vectors* and they play the main role in designing the SVMs. What characterizes a support vector over a nonsupport vector is that removing any support vector will have a big impact on the direction and position of the hyperplane. The equality in the equation above occurs only if the vector $\boldsymbol{x}$ is a support vector.

Hence, the basic idea in the SVM method is to maximize the margin, which can be achieved by maximizing the distance between the hyperplane and the support vectors. Since the distance between the hyperplane and a point x is given by

$$z = \frac{\left| \boldsymbol{w}^T \boldsymbol{x} + b \right|}{\|\boldsymbol{w}\|}$$

Maximizing z is achieved by minimizing $\|\boldsymbol{w}\|$ and maximizing the bias b simultaneously.

To design the optimal hyperplane, we first assume that we have the training set $\{(x_1, y_1), (x_2, y_2), \ldots, (x_n, y_n)\}$. Then, we have to solve the following quadratic programming (QP) problem:

$$\min_{w \in R^m} \frac{1}{2} \|\boldsymbol{w}\|^2$$

subject to the n inequality constraints:

$$y_i \left(\boldsymbol{w}^T \boldsymbol{x}_i + b \right) \geq 1, \quad \forall i = 1, \ldots, n$$

The solution of the above QP problem is the saddle node of the Lagrangian, given by

$$L(\boldsymbol{w}, b, \alpha) = \frac{1}{2} \|\boldsymbol{w}\|^2 - \sum_{i=1}^{n} \alpha_i \left[y_i \left(\boldsymbol{w}^T \boldsymbol{x}_i + b - 1 \right) \right]$$

where $= (\alpha_1, \alpha_2, \ldots, \alpha_n) \in R^n$, with $\alpha_i \geq 0, \forall i = 1, \ldots, n$ is the vector of the Lagrange multipliers.

To find the optimal solution, we minimize the Lagrangian with respect to $\boldsymbol{w}$ and b, and maximize it with respect to α.

If $(\hat{w}, \hat{b}, \hat{\alpha})$ is the optimal solution for the above QP problem, then the Lagrangian at this point satisfies the following first order necessary conditions:

1. $\dfrac{\partial L}{\partial b}(\hat{w}, \hat{b}, \hat{\alpha}) = 0$

2. $\dfrac{\partial L}{\partial w}(\hat{w}, \hat{b}, \hat{\alpha}) = 0$

If we expand the Lagrangian, we obtain the following form:

$$L(w,b,\alpha) = \frac{1}{2}\|w\|^2 - \sum_{i=1}^{n}\alpha_i y_i w^T x_i - \sum_{i=1}^{n}\alpha_i y_i b + \sum_{i=1}^{n}\alpha_i$$

By imposing the following condition:

$$\frac{\partial L}{\partial b}(\hat{w}, \hat{b}, \hat{\alpha}) = 0$$

gives

$$\sum_{i=1}^{n}\hat{\alpha}_i y_i = 0$$

By imposing the following condition:

$$\frac{\partial L}{\partial w}(\hat{w}, \hat{b}, \hat{\alpha}) = 0$$

it gives

$$\hat{w} = \sum_{i=1}^{n}\hat{\alpha}_i y_i x_i$$

The above form of $\hat{w}$ indicates that the optimal weight is a linear combination of the training set (x_i, y_i), for $i = 1, \ldots, n$.

According to the Kuhn–Tucker theorem, at the optimal solution, only the equality constraints can have nonzero Lagrange multipliers, that is, if

$$\hat{\alpha}_i \left(y_i \left(\hat{w}^T x_i + b \right) - 1 \right) = 0$$

then, $\hat{\alpha}_i \neq 0$, and if

$$\hat{\alpha}_i \left(y_i \left(\hat{w}^T x_i + b \right) - 1 \right) > 0$$

then

$$\hat{\alpha}_i = 0$$

Supposing that S_v is the set of support vectors and x_k does not belong to S_v, then $y_k \left(w^T x_k + b \right) > 1$, and it is desired that the corresponding coefficient α_k must be zero. On the other hand, if $x_i \in S_v$, then $y_i \left(w^T x_i + b \right) - 1 = 0$, and it is desired that the corresponding α_i is nonzero.

Now, by substituting the forms of $\left(\hat{w}, \hat{b} \right)$ in the Lagrangian, it becomes

$$L\left(\hat{w}, \hat{b}, \alpha \right) = J(\alpha) = \sum_{i=1}^{n} \alpha_i - \frac{1}{2} \sum_{i=1}^{n} \alpha_i \alpha_j y_i y_j \left(x_i \cdot x_j \right)$$

The optimal choice for the Lagrange multipliers is obtained by maximizing $J(\alpha)$ subject to the constraints of nonnegativity of the Lagrange multipliers. That is to solve the QP problem:

$$\max \sum_{i=1}^{n} \alpha_i - \frac{1}{2} \sum_{i=1}^{n} \alpha_i \alpha_j y_i y_j \left(x_i \cdot x_j \right)$$

Subject to the constraints:

$$\alpha_i \geq 0$$

and

$$\sum_{i=1}^{n} \alpha_i y_i = 0$$

Since only the support vectors are involved in the process of computing the optimal Lagrange multipliers, the above optimization problem turns to

$$\max \sum_{i:x_i \in S_v} \alpha_i - \frac{1}{2} \sum_{i,j:x_i,x_j \in S_v} \alpha_i \alpha_j y_i y_j \left(x_i \cdot x_j \right)$$

Subject to the following constraints:

$$\alpha_i \geq 0$$

with $\alpha_i = 0$ if x_i is not a support vector, and

$$\sum_{i:x_i \in S} \alpha_i y_i = 0$$

If $\hat{\alpha} = \left[\hat{\alpha}_1, \ldots, \hat{\alpha}_n \right]$ is the vector of the optimal Lagrange multipliers, then

$$\hat{w} = \sum_{i:x_i \in S_v} \hat{\alpha}_i y_i x_i$$

and if x_k is any support vector, then

$$\hat{b} = y_k - \hat{w}^T x_k$$

Finally, if z is any unclassified feature vector, the following classification function:

$$f(z) = \text{sign} \left(\sum_{i:x_i \in S_v} \hat{\alpha}_i y_i x_i \cdot z + \hat{b} \right)$$

If we set $K(x_i, x_j) = x_i \cdot x_j$, the term $\hat{w} \cdot x$ is written as

$$\hat{w} \cdot x = \sum_{i:x_i \in S_v} \hat{\alpha}_i y_i K(x_i, x)$$

and the functional $K(x_i, x) = x_i \cdot x$ is referred to as the *linear kernel*.

Now, if $|S| = s$, and $\{x_1, x_2, \ldots, x_s\}$ is the set of support vectors from the two classes C_1 and C_2. It is obvious that

$$\hat{w} \cdot x_k + b = y_k$$

or equivalently,

$$\sum_{j=1}^{s} \widehat{\alpha_j y_j} (x_j \cdot x_k) + b = y_k$$

That is,

$$\sum_{j=1}^{s} \widehat{\alpha_j y_j} K (x_j \cdot x_k) + b = y_k$$

We see that s linear equations in the $s+1$ unknowns $\alpha_1, \alpha_2, \ldots, \alpha_s$ and b of the above form can be formulated. In addition to those s equations, the equation

$$y_1 \alpha_1 + \cdots + y_s \alpha_s = 0$$

completes the set of equations into $s+1$ equations in $s+1$ unknowns.

Let $M \in R^{(s+1) \times (s+1)}$, $b \in R^{(s+1)}$, and $a \in R^{(s+1)}$ be defined as follows:

$$M = \begin{bmatrix} y_1 K(x_1, x_1) & y_2 K(x_2, x_1) & \cdots & y_s K(x_s, x_1) & 1 \\ \vdots & \vdots & \ddots & \vdots & \vdots \\ y_1 K(x_1, x_s) & y_2 K(x_2, x_s) & \cdots & y_s K(x_s, x_s) & 1 \\ y_1 & y_2 & \cdots & y_s & 0 \end{bmatrix},$$

$$a = \begin{bmatrix} \alpha_1 \\ \vdots \\ \alpha_s \\ b \end{bmatrix}, b = \begin{bmatrix} y_1 \\ \vdots \\ y_s \\ 0 \end{bmatrix}$$

The optimal weight $\hat{w}$ and the optimal bias $\hat{b}$ can be obtained by solving the linear system:

$$Ma = b$$

There are other kinds of kernels that are nonlinear and very efficient in manipulating the nonlinearly separable data.

8.2.2 The Case of Nonlinear Kernel

The kernel methods aim at generating a smooth separating nonlinear decision boundary, as can be seen in Figure 8.5.

The most popular kernels are the radial basis functions (RBFs) kernels. The Gaussian kernel is of the following form:

$$K\left(x_i, x_j\right) = e^{-\left(\|x_i - x_j\|^2 / \sigma^2\right)}$$

The Gaussian kernel has the property that $0 \le K(x_i, x_j) \le 1$. If the distance between x_i and x_j is very small (their features are close to each other), the value of the kernel remains close to 1. As the distance between them increases (their features are not so similar), the value of the kernel goes toward 0 (Figure 8.6).

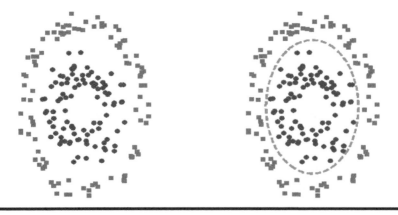

Figure 8.5 Nonlinearly separable data.

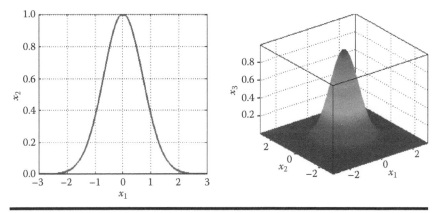

Figure 8.6 **The Gaussian RBF in two-dimensional feature space (left) and three-dimensional feature space (right).**

8.3 The SVM in MATLAB®

MATLAB deals with the data classification with the SVM through two main functions, namely, the "svmtrain" and the "svmclassify" functions.

The svmtrain function receives the set of feature vectors, the set of the corresponding classes, and the kind of the kernel (linear, radial basis, etc). The result from applying the svmtrain is a structure of the svm. The svm structure contains the fields of support vectors, optimal Lagrange multipliers, the bias, and the kind of the used kernel.

From the Iris database, we picked all the *Iris-setosa* and the *Irisversicolor* instances, where the classs of *Iris-setosa* is labeled as 1, whereas the class of the *Irisversicolor* is labeled −1. The number of all instances was 100. The SVM is trained with 80% of the data and 20% are used as testing dataset. The feature vectors are loaded into a matrix "Features" and the corresponding classes are loaded in a vector "Classes." The linear kernel function is used at the training phase. The testing instances are loaded in a matrix "TestingData," which is passed to the svmclassifier that was generated in the training phase. All the testing instances are classified correctly.

```
SVMModel = svmtrain(Features, Classes, 'Kernel_
Function', 'Linear');
NewClasses = svmclassify(SVMStruct, TestingData);
```

Running the above MATLAB script gives the following results:

```
SVMStruct =
            SupportVectors: [4 × 4 double]
                     Alpha: [4 × 1 double]
                      Bias: -0.0345
            KernelFunction: @linear_kernel
        KernelFunctionArgs: {}
                GroupNames: [80 × 1 double]
       SupportVectorIndices: [4 × 1 double]
                 ScaleData: [1 × 1 struct]
              FigureHandles: []
```

References

1. Ying, C. C. Y. *Learning with Support Vector Machines*. San Rafael, CA: Morgan and Claypool, 2011.
2. Jordan, M., Kleinberg, J., and Scholkopf, B. *Support Vector Machines*, Information Science and Statistics. New York: Springer, 2008.
3. Manning, C. D., Raghavan, P., and Schutze, H. *Introduction to Information Retrieval*. New York: Cambridge University Press, 2008.
4. Vapnik, V. N. *The Nature of Statistical Learning Theory*, 2nd Ed. New York: Springer-Verlag, 1999.

UNSUPERVISED LEARNING ALGORITHMS

Introduction

In Section II, we will discuss the following algorithms:

1. *K*-means
2. Gaussian mixture model
3. Hidden Markov model
4. Principal component analysis in context of dimensionality reduction

Chapter 9

k-Means Clustering

9.1 Introduction

The method of *k*-means clustering is one of vector quantization, originally from signal processing, which is popular for cluster analysis in data mining. This method of *k*-means clustering aims to partition n observations into k clusters in which each observation belongs to the cluster with the nearest mean, serving as a prototype of the cluster. This results in a partitioning of the data space into Voronoi cells.

The problem is computationally difficult (NP-hard); however, there are efficient heuristic algorithms that are commonly employed and converge quickly to a local optimum. These are usually similar to the expectation-maximization algorithm for mixtures of Gaussian distributions via an iterative refinement approach employed by both algorithms. Additionally, they both use cluster centers to model the data; however, *k*-means clustering tends to find clusters of comparable spatial extent, while the expectation-maximization mechanism allows clusters to have different shapes.

The algorithm has a loose relationship to the *k*-nearest neighbors (*k*-NN) classifier, a popular machine learning technique for classification that is often confused with *k*-means because of the *k* in the name. One can apply the 1-nearest neighbor classifier on the cluster centers obtained by *k*-means to classify new data into the existing clusters. This is known as *nearest centroid classifier* or *Rocchio algorithm*.

9.2 Description of the Method

Given data $\{x_1, x_2, ..., x_N\}$, where $x_j = (a_{j1}, a_{j2}, ..., a_{jm}) \in R^m$: $j = 1, ..., N$. The aim is to group the data into K clusters $C_1, C_2, ..., C_K$, centered at mean points $\mu_1, \mu_2, ..., \mu_K$, where $\mu_j \in R^m$, $j \in \{1, ..., K\}$. The desired clustering process must satisfy the two conditions:

1. Each point $x_i, i \in \{1, ..., N\}$ must belong to exactly one cluster $C_j, j \in \{1,...,K\}$ centered at point $\mu_j, j \in \{1,...,K\}$. In addition, x_i is assigned to a cluster C_j if and only if $\|x_i - \mu_j\| \leq \|x_i - \mu_k\|$ for $k \in \{1, ..., K\}$
2. The choice of the central points $\mu_1, ..., \mu_K$ is such that the functional

$$J(\mu_1, ..., \mu_K) = \sum_{k=1}^{K} \sum_{x_j \in C_k} \|x_j - \mu_k\|^2$$

   is minimum

In a cluster $C_j, j \in \{1,...,K\}$, the mean μ_j is computed through the rule

$$\mu_j = \frac{1}{N_j} \sum_{x_k \in C_j} x_k$$

where, N_j is the number of points in the cluster C_j and $N_1 + N_2 + \cdots + N_k = N$.

The *k*-means clustering method is an iterative method, which starts with a random selection of the *k*-means $\mu_1, \mu_2, ..., \mu_k$. Then, in each iteration, the data points are grouped in *k*-clusters, according to the closest mean to each of the points, and the mean is updated according to the points within the cluster. This process of grouping the data points according to the clusters means and updating the cluster means according to the set of points in the cluster continues until there is no further change in the clusters points, nor clusters means.

It is noteworthy that each time we run the *k*-means algorithm, it might give different means and clusters. This is because of the random selection of the initial *k*-means.

9.3 The *k*-Means Clustering Algorithm

The following algorithm receives a set of data points $\{x_1, \ldots, x_N\}$ and an integer K. It applies the *k*-mean algorithm and returns the cluster labels of the data points, the means, the within cluster sum of the distances of cluster points from their mean, and the distances of the points from the means.

```
Function K_MeansClustering({x₁,...,xₙ}, Integer K)
```
(1)   Set $j = 0$; $C_1 = C_2 = \cdots = C_K = \varphi$.
(2)   Select a random set of means $\{\mu_1^{(0)}, \mu_2^{(0)}, \ldots, \mu_k^{(0)}\}$.
(3)   At iteration j, for each $x_i, i = 1, \ldots, N$; find $k \in \{1, \ldots, K\}$ such that
$$\left\| x_i - \mu_k^{(j)} \right\| \leq \left\| x_i - \mu_l^{(j)} \right\|, \quad \forall l \in \{1, \ldots, K\}, \text{ and set}$$
$C_k = C_k \cup \{x_i\}$
(4)   Set $j = j + 1$
(5)   Within each cluster $C_k, k = 1, \ldots, K$; set $N_k = |C_k|$ and recompute new means $\{\mu_1^{(j)}, \mu_2^{(j)}, \ldots, \mu_k^{(j)}\}$, where
$$\mu_k^{(j)} = (1/N_k) \sum_{x_i \in C_k} x_i, \quad k = 1, \ldots, K$$
(6)   Repeat steps (3)-(5) until convergence.
(7)   Set $\mu_k = \mu_k^{(j)}$, $k = 1, \ldots, k$
(8)   for each $x_i, i = 1, \ldots, N$; find $k \in \{1, \ldots, K\}$, such that $x_i \in C_k$ and label x_i by k; that is $L_i = k$
(9)   for each $i \in 1, \ldots, N$ and $k = 1, \ldots, K$, set $\text{Dist}_{ik} = \| x_i - \mu_k \|$

(10)   Within each cluster $C_k, k = 1, ..., K$, compute
$$\text{Sum}_k = \sum_{x_i \in C_k} \| x_i - \mu_k \|$$

(11)   Return $\{L, \mu, \text{Sum}, \text{Dist}\}$

9.4 The *k*-Means Clustering in MATLAB®

The MATLAB's function `kmean` receives two parameters D and K, where D is an $N \times m$ matrix representing the unlabeled data, and K is a positive integer, representing the number of clusters. Then, the function `kmeans` returns and index vector **Idx** of type $N \times 1$, representing the data labels. The data labels contained in vector **Idx** are integers $1...K$, each represents the cluster to which the data belongs to. A second possible output of the *k*-means function are the matrix of clusters means μ of type $m \times K$, in which the elements of each column k are the coordinates of the mean μ_k, $k = 1...K$. Other outputs are the summations of data points from their means within each cluster, represented by a vector of type $K \times 1$, and the distances of the single points from their means within each cluster, represented by a matrix of type $N \times K$.

In this example, we consider the ecoli data consisting of 336 records. The data and its description are available at the University of California, Irvine, California, machine learning repository, website: https://archive.ics.uci.edu/ml/datasets/ Ecoli). The following MATLAB's code uses the *k*-means function to group the data into four clusters. It then plots the data and the clusters means.

In Figure 9.1. the plotted data and plotted means (in black) are shown.

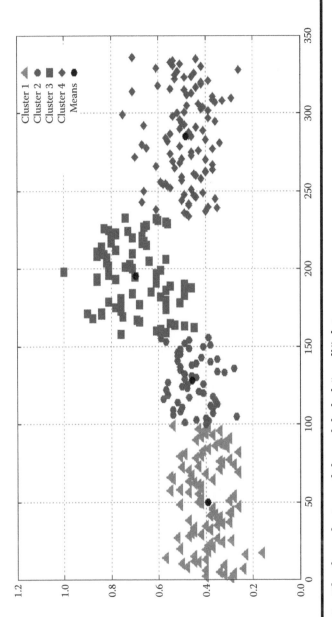

Figure 9.1 The four clusters of the unlabeled "ecoli" data.

```
clear; close all;
A = load('EColi1.txt'); % Reading data from file
EColi.txt
[m, n] = size(A);        % returning the size of the
file
B = A(1:end, 2:end-1);  % ignoring the data labels
(7 labels)
Idx = kmeans(B, 4);      % grouping the data into
four clusters
B1 = B(Idx==1);          % All data belonging to
cluster 1 are grouped in B1
B2 = B(Idx==2);          % All data belonging to
cluster 2 are grouped in B2
B3 = B(Idx==3);          % All data belonging to
cluster 3 are grouped in B3
B4 = B(Idx==4);          % All data belonging to
cluster 4 are grouped in B4

mu1 = mean(B1); % computing mean of cluster 1
mu2 = mean(B2); % computing mean of cluster 2
mu3 = mean(B3); % computing mean of cluster 3
mu4 = mean(B4); % computing mean of cluster 4

S1 = 1:length(B1);
% data in cluster 1 take the ranks from 1 to size B1
S2 = S1(end)+1:S1(end)+length(B2);
% data in cluster 2 take next indices to data in B1
S3 = S2(end)+1:S2(end)+length(B3);
% data in cluster 3 take next indices to data in B2
S4 = S3(end)+1:S3(end)+length(B4);
% data in cluster 4 take next indices to data in B3
plot(S1, B1, 'bo', S2, B2, 'rd', S3, B3, 'ms', S4,
B4, 'g^');
% plotting data
grid on;
hold on;
c1 = mean(S1); c2 = mean(S2); c3 = mean(S3); c4 =
mean(S4);
%computing mean indices for each cluster
plot(c1,mu1,'ko', c2,mu2,'kd', c3,mu3,'ks',
c4,mu4,'k^', 'LineWidth', 3);
% computing clusters means and plotting them.
```

Chapter 10

Gaussian Mixture Model

10.1 Introduction

As we know that each Gaussian is represented by a combination of mean and variance, if we have a mixture of M Gaussian distributions, then the *weight* of each Gaussian will be a third parameter related to each Gaussian distribution in a Gaussian mixture model (GMM). The following equation represents a GMM with M components.

$$p\left(x|\theta\right) = \sum_{k=1}^{M} w_k p(x \mid \theta_k)$$

where w_k represents the weight of the kth component. The mean and covariance of kth components are represented by $\theta_k = \left(\mu_k, \Sigma_k\right)$. $p(x|\theta_k)$, which is the Gaussian density of the kth component and is a D-variate Gaussian function of the following form:

$$p\left(x|\theta_k\right) \text{ or } p\left(x \middle| \left(\mu_k, \sum_k\right)\right) = \frac{1}{2\pi^{D/2} \left|\sum_k\right|^{1/2}} e^{\left\{-(1/2)(x-\mu_k)'\sum_k^{-1}(x-\mu_k)\right\}}$$

Sum of values of w_k for different values of k should not exceed 1 or $\sum_{k=1}^{M} w_k = 1$ and $w_k > 0$, $\forall k$.

When one is performing clustering using GMM, the goal is to find the model parameters (mean and covariance of each distribution as well as the weights), so that the resulting model best fits the data. In order to best fit the data, we should maximize the likelihood of the data given in the GMM model. This target can be achieved by using iterative expectation maximization (EM) algorithm, but initial estimates are required to execute this algorithm. If these initial estimates are poor, the algorithm can get stuck in local optima. A solution to this problem is to start with k-means and use the discovered mean and covariances of clusters as input for the EM algorithm.

Once we are able to fit the mixture model, we can explore the clusters by computing the posterior probability of data instances using each mixture component. GMM will assign each instance to a cluster based on calculated likelihood.

GMM is used in a number of applications, including speaker identification [1] and biometric verification [2].

10.2 Learning the Concept by Example

In order to understand how the GMMs can be used for clustering purpose, we will describe the process of clustering with one-dimensional GMM with three components. Each of the components has its respective mean and variance values. The graph (bell curve) of the associated probability density of each component has a peak at the mean. For our example we will have three bell curves with three different peaks. The data associated with each component is given in Table 10.1.

In MATLAB®, the relevant code to manage the above information is simple and is described as follows:

Table 10.1   The Means and Variances of Three Gaussian Distributions

Gaussian Component	Mean	Variance
1	−1	2.25
2	0	1
3	3	0.25

```
mu1 = [-1];
mu2 = [0];
mu3 = [3];
sigma1 = [2.25];
sigma2 = [1]
sigma3 = [.25];
```

We will randomly generate values from the three distributions in different proportion. For example, 30% of values will come from the first distribution, 50% of data will be from the second distribution, and 20% of data will be taken from the third distribution. For the generation of a sample of 1000 random values, the following MATLAB code will serve the purpose.

```
weight1 = [.3];
weight2 = [.5];
weight3 = [.2];
component_1 = mvnrnd(mu1,sigma1,300);
component_2 = mvnrnd(mu2,sigma2,500);
component_3 = mvnrnd(mu3,sigma3,200);
X = [component_1; component_2; component_3];
```

In order to understand how the distribution of three components looks like, we will plot the three distributions in one graph. The following MATLAB code will perform the required job.

```
gd1 = exp(- 0.5 * ((component_1 - mu1) / sigma1).^
2) / (sigma1 * sqrt(2 * pi));
gd2 = exp(- 0.5 * ((component_2 - mu2) / sigma2).^
2) / (sigma2 * sqrt(2 * pi));
```

```
gd3 = exp(- 0.5 * ((component_3 - mu3) / sigma3).^
2) / (sigma3 * sqrt(2 * pi));
plot(component_1,gd1,'.')
hold on
plot(component_2,gd2,'.')
hold on
plot(component_3,gd3,'.')
grid on

title('Bell Curves of three components')

xlabel('Randomly produced numbers')

ylabel('Gauss Distribution')
```

The result of the above code in shown in Figure 10.1.

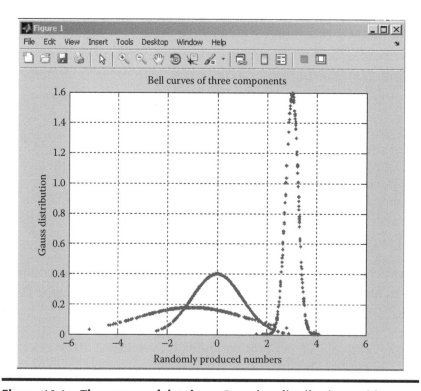

Figure 10.1 The curves of the three Gaussian distributions with means and variances as in Table 10.1.

In order to find the model that fits the three distributions, we can use the following MATLAB code:

```
gm1 = gmdistribution.fit (X,3);
a = pdf (gm1,X)
plot(X, a,'.')
```

This will result in the mixed distribution shown in Figure 10.2.

In order to find which points belong to which cluster, one single line of MATLAB code will suffice.

```
idx = cluster (gm1,X);
```

Since we are using 1000 random numbers from three different distributions that are present in sequential manner in 1000 × 1 array, *idx* will also be a one-dimensional array

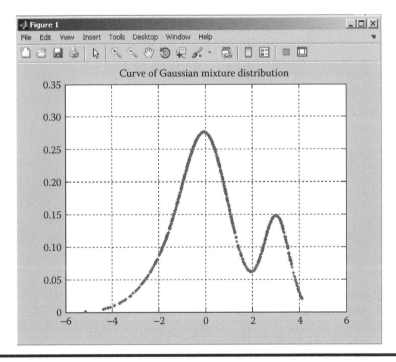

Figure 10.2   The mixed Gaussian distribution of the three Gaussian curves.

(1000 × 1) with each value having three possible values that are 1, 2, and 3, describing the cluster number to which the particular value belongs to. Since our pool of random numbers belonging to the three distributions were inserted sequentially, we expect that in variable idx there will be a long series of 300 1s followed by 500 2s and then 200 3s. We will plot idx in one dimension to understand the allocation of points to clusters. The following code will fulfill the requirement:

```
hold on;
for i = 1: 1000
    if idx(i) == 1
            plot(X(i),0,'r*')
    elseif idx(i) == 2
            plot(X(i),0,'b+')
    else
            plot(X(i),0,'go')
    end
end
title('Plot illustrating the cluster assignment');
ylim([-0.2 0.2]);
hold off
```

The result of the above code is shown in Figure 10.3.

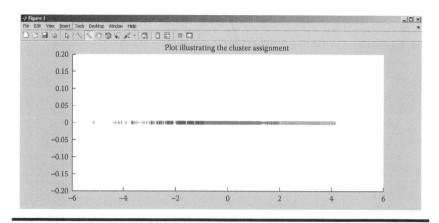

Figure 10.3   The three clusters corresponding to the three Gaussian distributions.

References

1. Reynolds, D. Speaker identification and verification using Gaussian mixture speaker models, *Speech Communication*, vol. 17, 91–108, 1995.
2. Stylianou, Y., Pantazis, Y., Calderero, F., Larroy, P., Severin, F., Schimke, S., Bonal, R., Matta, F., and Valsamakis, A. GMM—Based Multimodal Biometric Verification. Final Project Report 1, Enterface'05, Mons, Belgium, July 18–August 12, 2005.

Chapter 11

Hidden Markov Model

11.1 Introduction

A hidden Markov model (HMM) has invisible or unobservable states, but a visit to these hidden states results in the recording of observation that is a probabilistic function of the state. Given a sequence of observations, the problem is to infer the dynamical system that had produced the given sequence. This inference will result in a model for the underlying process. The three basic problems of HMMs are given below:

1. *Scoring problem*: The target is to find the *probability of an observed sequence* with HMM already given.
2. *Alignment problem*: Given a particular HMM, the target is to determine from an observation sequence the most likely sequence of underlying hidden states that might have generated it.
3. *Training problem*: The target is to create an HMM given a particular set of related training sequences.

Speech [1], handwriting, and gesture [2] recognitions are the areas of field of pattern recognition where the HMM is applied successfully. It is also used in part-of-speech tagging [3] and bioinformatics.

Section 11.2 provides practical examples of HMM, which belongs to the category of alignment problem.

11.2 Example

Consider the example of an employee heavily influenced by what is going on at his office. The impact of his boss's attitude toward him affects his attitude at home. The wife of this employee tries to understand the behavior of her husband in terms of what would have happened to him at office but she has little information about the parameters of her husband's office. Now, consider the following hypothetical situation.

The employee has a boss who has only one of the two attitudes toward his employees in every business day: (1) angry and (2) happy.

The employee has one of three possible reactions at his home: (1) keep silent, (2) constantly yell, and (3) talk happily.

We further suppose that the wife of the employee is good at machine learning and with her knowledge about different parameters, she decided to use HMM to guess what the attitude of the boss toward her husband on a particular series of days would be.

She formulated the following representation to make her scientific guess.

States:

  1. Angry
  2. Happy

Observations on four consecutive days:

Monday	Tuesday	Wednesday	Thursday
Silent	Yelling	Joking	Silent

(*Continued*)

Start probability:

1. Boss is angry: 0.4
2. Boss is happy: 0.6

Transition probability of states:

Probability that one day boss is angry and on the next day, he is still angry: 0.6

Probability that one day boss is angry and on the next day, he is happy: 0.4

Probability that one day boss is happy and on the next day, he is angry: 0.3

Probability that one day boss is happy and on the next day, he is still happy: 0.7

Emission Probabilities Representing Relationship between States and Observation	
Boss State: Angry	*Boss State: Happy*
Probability that employee is silent at home = 0.3	Probability that employee is silent at home = 0.4
Probability that employee is yelling at home = 0.6	Probability that employee is yelling at home = 0.1
Probability that employee is joking at home = 0.1	Probability that employee is joking at home = 0.5

```
start_probability = {'Rainy': 0.6, 'Sunny':
0.4}
transition_probability = {
'Rainy': {'Rainy': 0.7, 'Sunny': 0.3},
'Sunny': {'Rainy': 0.4, 'Sunny': 0.6},
}
emission_probability = {
'Rainy': {'walk': 0.1, 'shop': 0.4, 'clean': 0.5},
'Sunny': {'walk': 0.6, 'shop': 0.3, 'clean': 0.1},
}
```

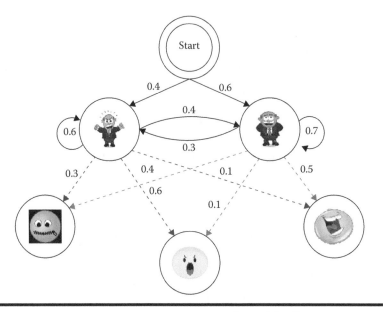

Figure 11.1 HMM parameters represented graphically.

The above information can be represented with the following coding in MATLAB®.

The above HMM parameters can be represented graphically as follows (Figure 11.1).

The wife's target is to discover possible boss's attitude with her husband. We will present the MATLAB code along with calculations. In order to keep things simple, we will avoid using loops. This means that the number of variable usage will increase and a code that is not generic but is valid for a problem having the same number of parameters. It is because the intention is to make the reader understand things quickly and easily.

11.3 MATLAB Code

```
states = {'angry', 'happy'};

%    Silent, Yelling, Joking, Silent. Silent = 1,
     Yelling =2, Joking=3
obs = [1,2,3,1];
```

```
%           Angry Happy
start_p = [0.4,0.6];

%           Angry Happy
trans_p = [0.6, 0.4; %transitions from Angry state
           0.3, 0.7]; %transitions from Happy state

%      Silent Yelling Joking
emit_p = [0.3,0.6,0.1; %emissions from Angry state
          0.4,0.1,0.5]; %emissions from Happy state

V = {};
V{1} = {start_p(1),states(1)}; % V(1) = (0.4,
'angry');
V{2} = {start_p(2),states(1)}; % V(2) = (0.6, 'happy');

output = 1; % On the first day, when employee was silent
    U = {};
    nextState = 1; %With observation = "silent", next
state = "angry"
        argmax = [];
        valmax = 0;
        sourceState = 1; % Observation = "Silent",next
state = "angry", source state = "angry"
            Vi = V{sourceState};
            v_prob = Vi{1}; v_path = Vi{2};
            p = emit_p(sourceState, obs(output)) *
trans_p(sourceState, nextState);
            v_prob = v_prob*p;
            if v_prob > valmax
                argmax = [v_path, states(nextState)];
                valmax = v_prob;
            end
        sourceState = 2; % Observation = "Silent", next
state = "angry", source state = "happy"
            Vi = V{sourceState};
            v_prob = Vi{1}; v_path = Vi{2};
            p = emit_p(sourceState, obs(output)) *
trans_p(sourceState, nextState);
            v_prob = v_prob*p;
            if v_prob > valmax
                argmax = [v_path, states(nextState)];
                valmax = v_prob;
            end
        U{nextState} = {valmax, argmax};
```

```
    nextState = 2; %With observation = "silent", next
state = "happy"
        argmax = [];
        valmax = 0;
        sourceState = 1; % Observation = "Silent", next
state = "happy", source state = "angry"
            Vi = V{sourceState};
            v_prob = Vi{1}; v_path = Vi{2};
            p = emit_p(sourceState, obs(output)) *
trans_p(sourceState, nextState);
            v_prob = v_prob*p;
            if v_prob > valmax
                argmax = [v_path, states(nextState)];
                valmax = v_prob;
            end
        sourceState = 2; % Observation = "Silent", next
state = "happy", source state = "happy"
            Vi = V{sourceState};
            v_prob = Vi{1}; v_path = Vi{2};
            p = emit_p(sourceState, obs(output)) *
trans_p(sourceState, nextState);
            v_prob = v_prob*p;
            if v_prob > valmax
                argmax = [v_path, states(nextState)];
                valmax = v_prob;
            end
    U{nextState} = {valmax, argmax};
    V = U;
```

The full code is available in HMM_for_book.m file.

The solution yielded by the program is given in the following diagram.

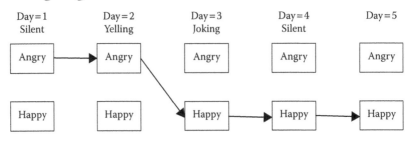

Rather than describing every step in detail, the following table is given to describe the result of the full code:

Output	next_state	Source_state	trans_p	emit_p	p	v_prob	v-prob (new)	Competing paths	Argmax	val.max
1{Silent}	Angry	Angry	0.6	0.3	0.18	0.4	0.072	{angry, angry}	{angry, angry}	0.072
		Happy	0.3	0.4	0.12	0.6	0.072	{happy, angry}	{angry, angry}	0.072
	Happy	Angry	0.4	0.3	0.12	0.4	0.048	{angry, happy}	{angry, happy}	0.048
		Happy	0.7	0.4	0.28	0.6	0.168	{happy, happy}	{happy, happy}	0.168
2{Yelling}	Angry	Angry	0.6	0.6	0.36	0.072	0.02592	{angry, angry, angry}	{angry, angry, angry}	0.02592
		Happy	0.3	0.1	0.03	0.168	0.00504	{happy, happy, angry}	{angry,angry,angry}	0.02592
	Happy	Angry	0.4	0.6	0.24	0.072	0.01728	{angry,angry,happy}	{angry,angry,happy}	0.01728
		Happy	0.7	0.1	0.07	0.168	0.01176	{happy, happy,happy}	{angry,angry,happy}	0.01728
3{Joking}	Angry	Angry	0.6	0.1	0.06	0.02592	0.001555	{angry,angry,angry,angry}	{angry,angry,angry,angry}	0.001555
		Happy	0.3	0.5	0.15	0.01728	0.002592	{angry,angry,happy,angry}	{angry,angry,happy,angry}	0.002592
	Happy	Angry	0.4	0.1	0.04	0.02592	0.001037	{angry, angry,angry,happy}	{angry,angry,angry,happy}	0.001037
		Happy	0.7	0.5	0.35	0.01728	0.006048	{angry,angry,happy,happy}	{angry,angry,happy,happy}	0.006048
1{Silent}	Angry	Angry	0.6	0.3	0.18	0.002592	0.000467	{angry,angry,happy,angry,angry}	{angry,angry,happy,angry,angry}	0.000467
		Happy	0.3	0.4	0.12	0.006048	0.000726	{angry, angry, happy,happy, angry}	{angry,angry,happy,happy,angry}	0.000726
	Happy	Angry	0.4	0.3	0.12	0.002592	0.000311	{angry,angry,happy,angry,happy}	{angry,angry,happy,angry,happy}	0.000311
		Happy	0.7	0.4	0.28	0.006048	0.001693	{angry,angry,happy,happy,happy}	{angry,angry,happy,happy,happy}	0.001693

As can be seen from the above table, following calculations and comparisons took place.

For day = 1 and next state = "Angry," two paths competed and since both had the same probability, the winner was (due to code preference) the first choice that is {Angry, Angry} with probability of 0.72.

For day = 1 and next state = "Happy," the competing paths were {angry, happy} and {happy, happy} since the probability of {happy, happy}(0.168) was higher than {angry, happy}; therefore, the path was selected for next usage.

From the table above, it can be seen that two paths selected for day 1 were used for subsequent calculation in day 2 and the two paths that were selected for day 2 were used in day 3 calculation.

Finally we were left with two choices:

1. Angry, angry, happy, happy, angry probability .000726
2. Angry, angry, happy, happy, happy … probability .000169

References

1. Rabiner, L. R. A tutorial on hidden Markov models and selected applications in speech recognition, *Proceedings of IEEE*, vol. 77, No. 2, 257–286, February 1989.
2. Starner, T. and Pentland, A. Real-time American sign language recognition from video using hidden Markov models, *Proceedings of the International Symposium on Computer Vision*, Coral Gables, FL. Los Alamitos, CA: IEEE CS Press, 1995.
3. Kupiec, J. Robust part-of-speech tagging using a hidden Markov model, *Computer Speech and Language*, vol. 6, 225–242, 1992.

Chapter 12

Principal Component Analysis

12.1 Introduction

Principal component analysis (PCA) is a statistical procedure that uses an orthogonal transformation to convert a set of observations of possibly correlated variables into a set of values of linearly uncorrelated variables called *principal components*. The number of principal components is less than or equal to the number of original variables. This transformation is defined in such a way that the first principal component has the largest possible variance (i.e., accounts for as much of the variability in the data as possible), and each succeeding component in turn has the highest variance possible under the constraint that it is orthogonal to the preceding components. The resulting vectors are an uncorrelated orthogonal basis set. The principal components are orthogonal because they are the eigenvectors of the covariance matrix, which is symmetric. PCA is sensitive to the relative scaling of the original variables.

12.2 Description of the Problem

Given data $\{x_1, \ldots, x_N\}$ where each element x_j has m attributes $\{a_{j1}, \ldots, a_{jm}\}$ and $j = 1, \ldots, N$. While applying a machine learning classification or clustering algorithm, it is not expected from all the features (attributes) to have valuable contributions to the resulting classification or clustering model. The attributes with the least contributions do act as noise, and reducing them from the data will result in a more effective classification or clustering model.

The PCA is a multivariate statistical technique aiming at extracting the features that represent most of the information in the given data and eliminating the least features with least information. Therefore, the main purpose of using the PCA is to reduce the dimensionality of the data without seriously affecting the structure of the data.

When collecting real data, usually the random variables that represent the data attributes are expected to be highly correlated. The correlations between these random variables can always be seen in the covariance matrix. The variances of the random variables are found in the diagonal of the covariance matrix. The sum of the variances (diagonal elements of the covariance matrix) gives the *overall variability*. In Table 12.1, we show the covariance matrix of the "ecoli" data consisting of seven attributes.

The PCA works to replace the original random variables with other sets of orthonormal set of vectors called the *principal components*. The first principal component is desired to pass as much closer as possible to data points, and the projection of the data into the space spanned by the first component is the best projection over spaces spanned by other vectors in one dimension. The second principal component is orthogonal to the first principal component and the plain spanned by it together with the first component is the closest plane to the data. The third, fourth, and other such components feature at the same logic.

Table 12.1 The Covariance Matrix of the Ecoli Dataset, with Seven Random Variables Representing the Seven Data Attributes

	Mcg	Gvh	Lip	Chg	Aac	Alm1	Alm2
Mcg	0.0378	0.0131	0.0025	0.0004	0.0052	0.0166	0.0068
Gvh	0.0131	0.0219	0.0006	0.0001	0.0013	0.0055	−0.0037
Lip	0.0025	0.0006	0.0078	0.0008	0.0008	0.0018	−0.0011
Chg	0.0004	0.0001	0.0008	0.0007	−0.0001	0	−0.0003
Aac	0.0052	0.0013	0.0008	−0.0001	0.0149	0.0074	0.0065
Alm1	0.0166	0.0055	0.0018	0	0.0074	0.0464	0.0365
Alm2	0.0068	−0.0037	−0.0011	−0.0003	0.0065	0.0365	0.0437

In the new space with the same dimension as the original space, the covariance matrix is a diagonal matrix, and preserves the total variability of the original covariance matrix.

In Figure 12.1a, we generated two-dimensional (2D) random data. Then, we used the PCA to project the 2D data into the first component (Figure 12.1b). We show the projected data in the figure.

12.3 The Idea behind the PCA

Given an m-dimensional data D of the form:

$$D = \begin{bmatrix} x_1 \\ x_2 \\ \vdots \\ x_N \end{bmatrix} = \begin{bmatrix} a_{11} & a_{12} & \cdots & a_{1m} \\ a_{21} & a_{22} & \cdots & a_{2m} \\ \vdots & \vdots & \ddots & \vdots \\ a_{N1} & a_{N2} & \cdots & a_{Mm} \end{bmatrix}$$

The data D can be visualized by Figure 12.2. Hence, each data element x_j is represented by an m-dimensional row vector.

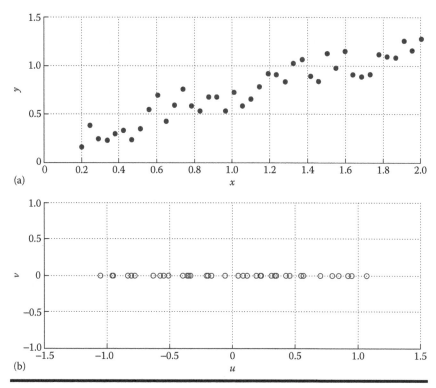

Figure 12.1 (a) The original two-dimensional data in the feature space and (b) the reduced data in one-dimensional space.

		Attributes			
		a_1	a_2	$\cdots$	a_m
Instances	x_1	a_{11}	a_{12}	$\cdots$	a_{1m}
	x_2	a_{21}	a_{22}	$\cdots$	a_{2m}
	$\vdots$	$\vdots$	$\vdots$	$\ddots$	$\vdots$
	x_N	a_{N1}	a_{N2}	$\cdots$	a_{Nm}

Figure 12.2 The shape of a dataset consisting of N feature vectors, each has m attributes.

The main purpose of using the PCA is to reduce the high-dimensional data with a dimension m into a lower dimensional data of dimension k, where $k \leq m$. The basis elements of the reduced space V constitute an orthonormal set. That is, if $\{v_1, v_2, \ldots, v_k\}$ is the basis of V, then:

$$v_i^T \cdot v_j = \begin{cases} 1, & j = i \\ 0, & j \neq i \end{cases}$$

The singular value decomposition (SVD) plays the main role in computing the basis elements $\{v_1, \ldots, v_k\}$.

12.3.1 The SVD and Dimensionality Reduction

Suppose that C is the covariance matrix that is extracted from a given dataset D. The elements of the covariance matrix are the covariances between the random variables representing the data attributes (or features). The variance of the random variables lies in the diagonal of C, and their sum is the total variability. Generally, the SVD works to express matrix C as a product of three matrices: U, Σ, and V. Because matrix C is an $m \times m$ matrix, the SVD components of matrix C are given by

$$C = U\Sigma V^T$$

where:
   U and V are orthogonal $m \times m$ matrices
   Σ is an $m \times m$ diagonal matrix

The diagonal elements of Σ are the eigenvalues of the covariance matrix C, ordered according to their magnitudes from bigger to smaller. The columns of U (or V) are the eigenvectors of matrix C that correspond to the eigenvalues of C.

By factoring matrix C into its SVD components, we would be rewarded with three benefits:

1. The SVD identifies the dimensions along which data points exhibit the most variation, and order the new dimensions accordingly. The total variation exhibited by the data is equal to the sum of all eigenvalues, which are the diagonal elements of the matrix Σ and the variance of the jth principal component is the jth eigenvalue.
2. Replace the correlated variables of the original data by a set of uncorrelated ones that are better exposed to the various relationships among the original data items. The columns of matrix U, which are the eigenvectors of C, define the principal components, which act as the new axes for the new space.
3. As a result of benefit (2), the SVD finds the best approximation of the original data points using fewer dimensions. Reducing the dimension is done through selecting the first k principal components, which are the columns of matrix U.

12.4 PCA Implementation

The steps to implement the PCA are as follows:

Step 1: Compute the mean of the data as follows:
Let

$$\bar{x} = \begin{bmatrix} \bar{a}_1 & \bar{a}_2 & \cdots & \bar{a}_m \end{bmatrix} = \frac{1}{N}\sum_{j=1}^{N} x_j$$

Step 2: Normalize the data by subtracting the mean value $\bar{x}$, from all the instances x_i giving the adjusted mean vectors $x_i - \bar{x}$. Therefore, the normalized data $\bar{D}$, is given by

$$\bar{D} = \begin{bmatrix} x_1 - \bar{x} \\ x_2 - \bar{x} \\ \vdots \\ x_N - \bar{x} \end{bmatrix}$$

The values $x_i - \bar{x}$ are the deviations of the data from its mean; hence, the adjusted mean features are centered at zero.

Step 3: Construct the covariance matrix $COV(\bar{D}) \in R^{m \times m}$ from $\bar{D}$ as follows:

$$\left(COV(\bar{D})\right)_{ij} = \frac{1}{N}\sum_{n=1}^{N}\left(x_{ni} - \bar{a}_i\right)\cdot\left(x_{nj} - \bar{a}_j\right)$$

Step 4: Let $\{\lambda_1, \lambda_2, \ldots, \lambda_m\}$ be the set of eigenvalues of the covariance matrix $COV(\bar{D})$, in a way that $|\lambda_1| \geq |\lambda_2| \geq \ldots \geq |\lambda_m|$ and $\{v_1, v_2, \ldots, v_m\}$ are the corresponding set of eigenvectors. The k vectors $v_1, v_2, \ldots, v_k$ represent the first k principal components.

Step 5: Construct the following matrix:

$$F = \begin{bmatrix} v_1 & v_2 & \cdots & v_k \end{bmatrix} \in R^{m \times k}$$

The reduced data in k-dimensions will be

$$\bar{D}_{reduced} = \bar{D} \cdot F \in R^{N \times k}$$

12.4.1 Number of Principal Components to Choose

When applying the PCA, the variances of the random variables are represented by the eigenvalues of the covariance

Table 12.2 Variances of the Seven Principal Components of Ecoli Data

K	1	2	3	4	5	6	7
Var(k)	0.5162	0.7604	0.8446	0.9187	0.9678	0.9962	1

matrix that are located at the main diagonal of matrix Σ. In matrix Σ the eigenvalues are sorted according to their magnitudes, from bigger to smaller. That means, the principal component with bigger variance comes first. Let $S \in R^m$ be the vector of the diagonal elements of Σ. The variance retained by choosing the first k principal components is given by

$$\frac{\sum_{j=1}^{k} S_j}{\sum_{j=1}^{m} S_j}$$

We applied the PCA algorithm to the "ecoli" data, and the variance retained by the first k principal components are explained in Table 12.2.

This means that if we select only the first principal component, then we can retain only 51.6% of the variance. If we select the first two principal components, then 76% of the variance will be retained, and so on.

By knowing this, we can determine how many principal components to select. For example, if we need to retain 90% at least of the variance, then we shall choose the first four principal components, if we need 99% of the variance, then we shall select the first six principal components.

12.4.2 Data Reconstruction Error

If $\tilde{x}_i = x_i - \bar{x} \in \bar{D}$, then $z_i = \tilde{x}_i \cdot F \in R^k$ is the projection of $\tilde{x}_i$ in the new linear space spanned by the orthonormal basis

Table 12.3 Reconstruction Errors Obtained by Applying PCA with *k* Principal Components

K	1	2	3	4	5	6	7
Error(k)	14.2171	4.9019	4.316	2.8588	1.6533	0.2207	0

vectors $\{v_1, v_2, \ldots, v_k\}$. Moving from R^m to R^k is reversible and one can reconstruct $\hat{x}_i$, which is an approximation to $\tilde{x}_i$, where:

$$\hat{x}_i = z_i \cdot F^T \in R^m$$

The error associated with the data reconstruction $\text{Error}_{\text{Rec}}$ is given by

$$\text{Error}_{\text{Rec}} = \sum_{i=1}^{N} \left\| \tilde{x}_i - \hat{x}_i \right\|^2$$

The reconstruction errors obtained by applying the PCA with *k* principal components are explained in the Table 12.3.

12.5 The Following MATLAB® Code Applies the PCA

```
function [Dred, Drec, Error] = PCAReduction(D, k)
% D is a dataset, consisting of N instances (rows)
and m features (columns)
% k is the new data dimension, where 0 <= k <= m
[N, m] = size(D) ; % Number of data instances N and
number of features is m
```

```
mu = mean(D) ;       % mean of data set --> mu

Db = zeros(size(D)) ; % Db is the mean adjusted
dataset
for j = 1 : N
    Db(j, :) = D(j, :) - mu ;
    % subtracting the mean from the N data
instances
end

C = zeros(m) ; % C is the covariance matrix
for i = 1 : m
    for j = 1 : m
        C(i, j) = 0 ;
        for n = 1 : N
            C(i, j) = C(i, j) + Db(n, i)*Db(n, j)/N ;
        end
    end
end

[U, S, V] = svd(C) ; % Applying the SVD to the
covariance matrix
F = V(:, 1:k) ;       % F is the matrix with the
first k components
Dred = Db*F ; % Generating the reduced data
Drec = Dred*F' ; % Reconstructing the mean adjusted
data in m-dimensions

for j = 1 : N
    Drec(j, :) = Drec(j, :) + mu ; % Reconstructing
the original data
end
Error = norm(Drec-D, 2)^2 % Computing the error
```

We applied the above algorithm to the Iris dataset with $k = 2$. In Figure 12.3, we show the plot of the reduced data (which is 2D).

We also plotted the data in a three-dimensional reduced space in Figure 12.4.

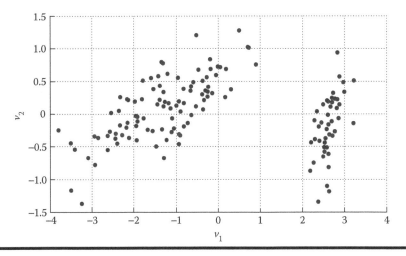

Figure 12.3   Two-dimensional reduced data.

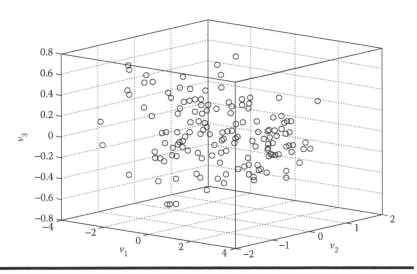

Figure 12.4   Three-dimensional reduced space.

12.6 Principal Component Methods in Weka

1. Open Weka and from the Weka explorer open the ecoli.
   arff file using the Open File button. The file ecoli.arff
   contains the ecoli dataset.

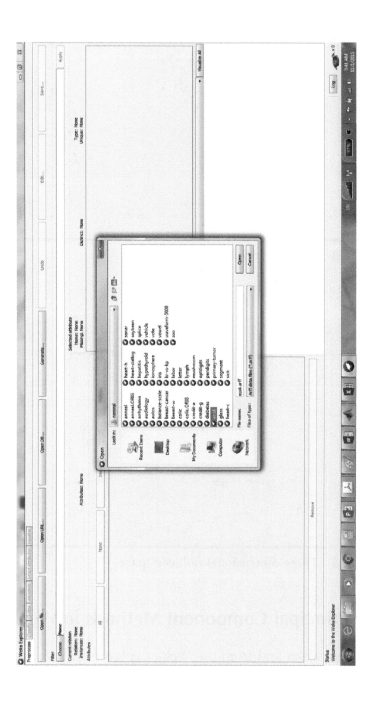

2. From the preprocess tab, choose Filter–>unsupervised–>
   attribute–>principal component.

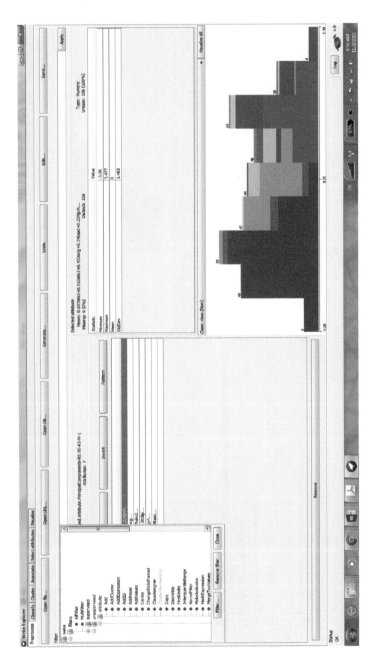

3. Press the apply button to enable the principal component filter to work.

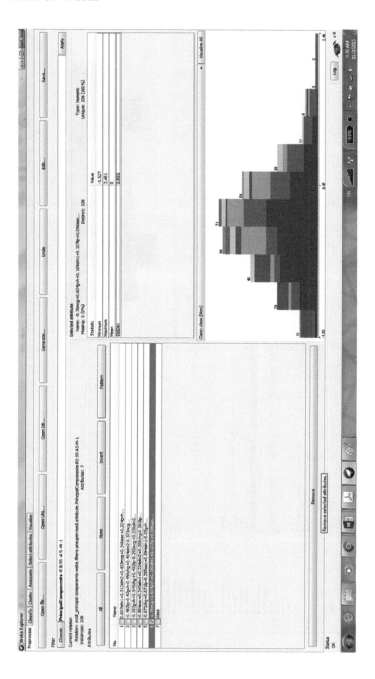

4. The statistics of the new resulting features, including the maximum and minimum values, the mean, and the standard deviation, can be seen.
5. You can notice that Weka has already reduced the number of features from seven to six. That is, the dimension is reduced by one.
6. However, Weka still enables the users to do further reductions if needed, where the standard deviation can be used to determine which features to eliminate and which features to keep. The smaller the standard deviation, higher will be the probability of that feature being eliminated.

12.7 Example: Polymorphic Worms Detection Using PCA

12.7.1 Introduction

Internet worms pose a major threat to the Internet infrastructure security, and their destruction causes loss of millions of dollars. Therefore, the networks must be protected as much as possible to avoid losses. This thesis proposes an accurate system for signature generation for Zero-day polymorphic worms.

This example consists of two parts.

In part one, polymorphic worm instances are collected by designing a novel double-honeynet system, which is able to detect new worms that have not been seen before. Unlimited honeynet outbound connections are introduced to collect all polymorphic worm instances. Therefore, this system produces accurate worm signatures. This part is out of the scope of this example.

In part two, signatures are generated for the polymorphic worms that are collected by the double-honeynet system.

Both a Modified Knuth–Morris–Pratt (MKMP) algorithm, which is string matching based, and a modified principal component analysis (MPCA), which is statistics based, are used. The MKMP algorithm compares the polymorphic worms' substrings to find the multiple invariant substrings that are shared between all polymorphic worm instances and uses them as signatures. The MPCA determines the most significant substrings that are shared between polymorphic worm instances and uses them as signatures [1].

12.7.2 SEA, MKMP, and PCA

This section discusses two parts. The first part presents our proposed substring exaction algorithm (SEA), MKMP algorithm, and MPCA, which are used to generate worm signatures from a collection of worm variants captured by our double-honeynet system [1].

To explain how our proposed algorithms generate signatures for polymorphic worms, we assume that we have a polymorphic worm A, which has n instances (A_1, A_2, ..., A_n). Generating a signature for polymorphic worm A involves two steps:

■ First, we generate the signature itself.
■ Second, we test the quality of the generated signature by using a mixed traffic (new variants of polymorphic worm A, and normal traffic).

Before stating the details of our contributions and the subsequent analysis part, we briefly mention an introduction about string matching search method and the original Knuth–Morris–Pratt (KMP) algorithm to give a clear picture of the subject topic.

The second part discusses the implementation results of our proposed algorithms.

12.7.3 Overview and Motivation for Using String Matching

In this and the following sections, we will describe SEA, MKMP algorithm, and modified PCA to highlight our contributions.

String matching [2] is an important subject in the wider domain of text processing. String matching algorithms are basic components used in implementations of practical software used in most of the available operating systems. Moreover, they emphasize programming methods that serve as paradigms in other fields of computer science (system or software design). Finally, they also play an important role in theoretical computer science by providing challenging problems.

String matching generally consists of finding a substring (called a *pattern*) within another string (called the *text*). The pattern is generally denoted as

$x = x[0..m-1]$

Whose length is m and the text is generally denoted as

$y = y[0..n-1]$

whose length is n. Both the strings pattern and text are built over a finite set of characters, which is called the *alphabet*, and is denoted by Σ whose size is denoted by σ.

The string matching algorithm plays an important role in network intrusion detection systems (IDSs), which can detect malicious attacks and protect the network systems. In fact, at the heart of almost every modern IDS, there is a string matching algorithm. This is a very crucial technique because it allows detection systems to base their actions on the content that is actually flowing to a machine. From a vast number of packets, the string identifies those packets that contain data, matching the fingerprint of a known attack. Essentially, the string matching algorithm compares the set of strings in the ruleset with the data seen in the packets, which flow across the network.

Our work uses SEA and MKMP algorithm (which are based on string matching algorithms) to generate signatures for polymorphic worm attacks. The SEA aims at extracting substrings from polymorphic worm, whereas the MKMP algorithm aims to find out multiple invariant substrings that are shared between polymorphic worm instances and to use them as signatures.

12.7.4 The KMP Algorithm

The KMP string searching algorithm [1] searches for occurrences of a *word*, *W*, within a main *text string*, *S*, by employing the observation that when a mismatch occurs, the word itself embodies sufficient information to determine where the next match could begin, thus bypassing reexamination of previously matched characters [1].

Let us take an example to illustrate how the algorithm works. To illustrate the algorithm's working method, we will go through a sample run (relatively artificial) of the algorithm. At any given time, the algorithm is in a state determined by two integers, *m* and *i*. The integer *m* denotes the position within *S*, which is the beginning of a prospective match for *W*, and *i* denotes the index in *W* denoting the character currently under consideration. This is depicted at the start of the run as follows:

```
m:  01234567890123456789012
S:  ABC ABCDAB ABCDABCDABDE
W:  ABCDABD
i:  0123456
```

We proceed by comparing successive characters of *W* to *parallel* positional characters of *S*, moving from one to the next if they match. However, in the fourth step in our noted case, we get that *S[3]* is a space and *W[3]* is equal to the character D (i.e., $W[3]$ = "D"), which is a mismatch. Rather than beginning to search again at the position S [1], we note that no A occurs between positions 0 and 3 in S except at 0. Hence, having checked all those characters previously, we know that there is no chance of finding the beginning of a match if we check them again. Therefore, we simply move on to the next character, setting $m = 4$ and $i = 0$.

```
m:  01234567890123456789012
S:  ABC ABCDAB ABCDABCDABDE
W:      ABCDABD
i:      0123456
```

We quickly obtain a nearly complete match "ABCDAB," but when at $W[6]$ ($S[10]$), we again have a discrepancy. However, just prior to the end of the current partial match, we passed an "AB" which could be the beginning of a new match, so we must take this into consideration. As we already know that these characters match the two characters prior to the current position, we need not check them again; we simply reset $m = 8$, $i = 2$, and continue matching the current character. Thus, not only do we omit previously matched characters of S but also previously matched characters of W.

```
m:  01234567890123456789012
S:  ABC ABCDAB ABCDABCDABDE
W:            ABCDABD
i:            0123456
```

We continue with the same method of matching till we match the word W.

12.7.5 Proposed SEA

In this subsection, we show how our proposed SEA is used to extract substrings from one of the polymorphic worm variants that are collected by the double-honeynet system.

This subsection and Section 12.7.6 show the signature generation process for polymorphic worm A using the SEA and the MKMP algorithm.

Let us assume that we have a polymorphic worm A, that has n instances $(A_1, \ldots, A_n)$ and A_i has length M_i, for $i = 1, \ldots, n$. Assume that A_1 is selected to be the instance from which we extract substrings and the A_1 string contains $a_1\, a_2\, a_3 \ldots a_{m1}$. Let X be the minimum length of a substring that we are going to extract from A_1. The first substring from A_1 with length X, is $(a_1\, a_2 \ldots a_x)$. Then, we shift one position to the right to extract a new substring, which will be $(a_2\, a_3 \ldots a_{x+1})$. Continuing this way, the last substring from A_1 will be $(a_{m1}-X + 1 \ldots a_{m1})$. In general, for instance, A_i has length equal to M, and if a minimum length of the substring that we are going to extract from A_1 equals to X, then the total number of substrings (TNSs) that will be extracted from A_i could be obtained by the following equation [1]:

```
TNS (Ai) = M-X+1
```

The next step is to increase X by one and start new substrings extraction from the beginning of A_1. The first substring will be $(a_1\, a_2 \ldots a_{x+1})$. The substrings extraction will continue satisfying this condition: $X < M$.

Figure 12.5 and Table 12.4 show all substrings extraction possibilities using the proposed SEA from the string ZYXCBA, assuming the minimum length of X is equal to three.

The output of the SEA will be used by both the MKMP algorithm and the modified PCA method. The MKMP algorithm uses the substrings extracted by the SEA to search the occurrences of each substring in the remaining of the instances $(A_2, A_3, \ldots, A_n)$. The substrings that occur in all the remaining instances will be considered as worm signature. To clarify some of the points noted here, we will present the details of the MKMP algorithm in Section 12.7.6 [1].

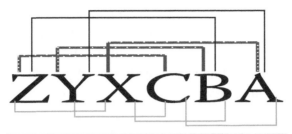

Light gray line $X = 3$. The substrings are ZYX, YXC, XCB, CBA	
Dark gray line $X = 4$. The substrings are ZYXC, YXCB, XCBA	
Black line $X = 5$. The substrings are ZYXCB, YXCBA	

Figure 12.5   Extraction substrings.

Table 12.4   Substrings Extraction

No. of Subtractions	Length of X	Substrings
S1,1	3	ZYX
S1,2	3	YXC
S1,3	3	XCB
S1,4	3	CBA
S1,5	4	ZYXC
S1,6	4	YXCB
S1,7	4	XCBA
S1,8	5	ZYXCB
S1,9	5	YXCBA

12.7.6 An MKMP Algorithm

Here, we describe our modification to the KMP algorithm. As mentioned in Section 12.7.4, the KMP algorithm searches for occurrences of W (word) within S (text string). Our modification of the KMP algorithm is to search for occurrence of different words (W_1, W_2, ..., W_n) within string text S. For example, say we have a polymorphic worm A, with

n instances $(A_1, A_2, \ldots, A_n)$. Let us select A_1 to be the instance from which we would extract substrings. If nine substrings are extracted from A_1, each substring will be W_i, for $i = 1$ to 9. This means that A_1 has nine words $(W_1, W_2, \ldots, W_9)$, whereas the remaining instances $(A_2, A_3, \ldots, A_n)$ are considered as S *text string* [1].

Considering the above example, the MKMP algorithm searches the occurrences of W_1 in the remaining instances of S $(A_2, A_3, \ldots, A_n)$. If W_1 occurs in all remaining instances of S, then we consider it as signature, otherwise we ignore it. The other words $(W_2, W_3, \ldots, W_9)$ are similarly dealt with. For example, if W_1, W_5, W_6, and W_9 occur in all remaining instances of S, then W_1, W_5, W_6, and W_9 are considered a signature of the polymorphic worm A.

12.7.6.1 Testing the Quality of the Generated Signature for Polymorphic Worm A

We test the quality of the generated signature for polymorphic worm A by using a mixed traffic (new variants of polymorphic worm A and normal traffic, i.e., innocuous packets). The new variants of polymorphic worm A are not the same variants that are used to generate the signature. Let us assume that our system received a packet P (where P contains either malicious or innocuous data). The MKMP algorithm compares P payload against the generated signature to determine whether P is a new variant of polymorphic worm A or not. The MKMP algorithm considers P as a new variant of the polymorphic worm A, if all the substrings of the generated signature appear in P.

12.7.7 A Modified Principal Component Analysis

12.7.7.1 Our Contributions in the PCA

In Sections 12.7.7.1 and 12.7.7.2, we show the signature generation process for polymorphic worm A using an MPCA.

In our work, instead of applying PCA directly, we have made appropriate modifications to fit it with our mechanism. Our contribution in the PCA method is in combining the PCA (i.e., extend) with the proposed SEA to get more accurate and relatively faster signatures for polymorphic worms. The extended method (SEA and PCA) is termed MPCA. We have previously mentioned that the polymorphic worm evades the IDSs by changing its payload in every infection attempts; however, there are some invariant substrings that will remain fixed (i.e., some substrings will not change) in all polymorphic worm variants, so the SEA extracts substrings from the polymorphic worm in a good way (i.e., it will extract all the possibilities of substrings from a polymorphic worm variant, which contain worm signature) that helps us to get accurate signatures. After the SEA extracts the substrings, it will pass those to the PCA, thus, easing the heavy burden to the PCA in terms of time (i.e., the PCA directly will start by determining the frequency Count of each substring in rest of the instances without doing substring extraction process).

After the PCA receives the substrings from SEA, it will determine the frequency count of each substring in the remaining instances $(A_2, A_3, ..., A_n)$. Finally, the PCA will determine the most significant data on the polymorphic worm instances and use them as signature []. We present the details in Section 12.7.7.1.1.

12.7.7.1.1 Determination of Frequency Counts

Here, we determine the frequency count of each substring S_i (A_1 substrings), in each of the remaining instances $(A_2, ..., A_n)$. Then, we apply the PCA on the frequency count data to reduce the dimension and get the most significant data.

12.7.7.1.2 Using PCA to Determine the Most Significant Data on Polymorphic Worm Instances

The methodology of employing PCA to the given problem is outlined below.

Let $\boldsymbol{F}_i$ denote the vector of frequencies $(F_{i1}, \ldots, F_{iN})$ of the substring S_i in the instances $(A_1, \ldots, A_n)$, $i = 1, \ldots, L$.

We construct the frequency matrix F by letting F_i be the ith row of **F**, provided that $\boldsymbol{F}_i$ is not the zero vector [3].

$$F = \begin{pmatrix} f_{11} & \cdots & f_{1N} \\ \vdots & \ddots & \vdots \\ f_{L1} & \cdots & f_{LN} \end{pmatrix}$$

12.7.7.1.3 Normalization of Data

The normalization of the data is applied by normalizing the data in each row of the matrix **F**, yielding a matrix **D** $(L \times N)$.

$$\mathbf{D} = \begin{pmatrix} d_{11} & \cdots & d_{1N} \\ \vdots & \ddots & \vdots \\ d_{L1} & \cdots & d_{LN} \end{pmatrix}$$

$$d_{ik} \leftarrow \frac{f_{ik}}{\sum_{j=1}^{N} f_{ij}}$$

12.7.7.1.4 Mean Adjusted Data

To get the data adjusted around zero mean, we use the following formula:

$$g_{ik} \leftarrow d_{ik} - \bar{d}_i \quad \forall i,k$$

where $\bar{d}_i$ is the mean of the ith vector

$$= \frac{1}{N} \sum_{j=1}^{N} d_{ij}$$

The data adjust matrix **G** is given by

$$
\mathbf{G} = \begin{pmatrix} g_{11} & \cdots & g_{1N} \\ \vdots & \ddots & \vdots \\ g_{L1} & \cdots & g_{LN} \end{pmatrix}
$$

12.7.7.1.5 Evaluation of the Covariance Matrix

Let g_i denote the ith row of **G**, then the covariance between any two vectors $\mathbf{g_i}$ and $\mathbf{g_j}$ is given by

$$
\mathrm{Cov}(\mathbf{g_i}, \mathbf{g_j}) = C_{ij} = \frac{\sum_{k=1}^{L} (d_{ik} - \bar{d}_i)(d_{jk} - \bar{d}_j)}{N-1}
$$

Then the covariance matrix C ($N \times N$) is given by

$$
\mathbf{C} = \begin{pmatrix} C_{11} & \cdots & C_{1N} \\ \vdots & \ddots & \vdots \\ C_{N1} & \cdots & C_{NN} \end{pmatrix}
$$

12.7.7.1.6 Eigenvalue Evaluation

Evaluate the eigenvalues of the matrix **C** from its characteristic polynomial $|C - \lambda I| = 0$, and then compute the corresponding eigenvectors.

12.7.7.1.7 Principal Component Evaluation

Let $L_1, L_2, \ldots, L_N$ be the eigenvalues of the matrix C obtained by solving the characteristic equation $|C - \lambda I| = 0$. If necessary, resort the eigenvalues of C in a descending order such that $|L_1| >= \cdots >= |L_N|$. Let $\mathbf{V_1}, \mathbf{V_2}, \ldots, \mathbf{V_N}$ be the eigenvectors of matrix

C corresponding to the eigenvalues L_1, L_2, ..., L_N. The k principal components are given by V_1, V_2, ..., V_K, where $K<=N$ [3].

12.7.7.1.8 Projection of Data Adjust along the Principal Component

Let V be the matrix that has the k principal components as its columns. That is,

$$V = [V_1, V_2, ..., V_K]$$

Then, feature descriptor (FD) is obtained from the following equation:

$$\text{Feature descriptor} = V^T \times F$$

To determine the threshold of polymorphic worm A, we use a distance function (Euclidean distance) to evaluate the maximum distance between the rows of F and the rows of FD. The maximum distance R works as a threshold [3].

12.7.7.2 Testing the Quality of Generated Signature for Polymorphic Worm A

In Section 12.7.7.1, we calculated the FD and the threshold for polymorphic worm A. Here, we test the quality of generated signature for polymorphic worm A by using a mixed traffic (new variants of polymorphic worm and normal traffic, i.e., innocuous packets), the new variants of polymorphic worm A are not the same variants that are used to generate the signature.

Let us assume that our system received a packet P (where P contains either malicious or innocuous data). The MPCA performs the following steps to determine whether P is a new variant of polymorphic worm A or not [1]:

■ Determine frequencies of the substrings of W array in P (W array contains extracted substrings of A_1, as mentioned earlier). This will produce a frequency matrix F_1.

■ Calculate the distance between the polymorphic worm FD and F_1 using Euclidean distance. This will produce a distance matrix D_1.

■ Compare the distances in D_1 to the threshold R of polymorphic worm A. If any <= the threshold, classify P as a new variant of polymorphic worm A.

12.7.7.3 Clustering Method for Different Types of Polymorphic Worms

When our network receives different types of polymorphic worms (mixed polymorphic worms), we must first separate them into clusters and then generate signatures to each cluster as the same as in Sections 12.7.7.1 and 12.7.7.2. To perform the clustering, we use Euclidean distance, which is the most familiar distance metric. Euclidean distance is frequently used as a measure of similarity in the nearest neighbor method []. Let $X = (X_1, X_2, ..., X_p)'$ and $Y = (Y_1, Y_2, ..., Y_p)'$. The Euclidean distance between X and Y is as follows [1]:

$$d(X,Y) = \sqrt{(X-Y)'(X-Y)}$$

12.7.8 Signature Generation Algorithms Pseudo-Codes

Here, we describe the signature generation algorithms pseudo-codes. These algorithms were discussed above, and as mentioned there, generating a signature for polymorphic worm A involves two steps:

■ First, we generate the signature itself.

■ Second, we test the quality of the generated signature by using a mixed traffic (new variants of polymorphic worm A and normal traffic).

12.7.8.1 Signature Generation Process

This section shows the pseudo-codes for generating a signature for polymorphic worm A using SEA, MKMPA, and MPCA.

12.7.8.1.1 SEA Pseudo-Code

In the following, we describe the pseudo-code of the SEA, which was discussed above. The goal of SEA is to extract substrings from the first instance of polymorphic worm *A* and then to put them in an array *W* [1].

SEA pseudo-code:

```
1. Function SubstringExtraction:
2. Input (a file A1: First instance of polymorphic
   worm A, x: minimum substring length)
3. Output: (W: array of substrings of A1 with a
   minimum substring length x)
4. Define variables:
   Integer M : Length of file A1
   Integer X: Maximum substring length
   Integer z: (x<=z<=X) takes the lengths x to X
   Integer Tz: Total number of substrings of
   file A1 with a substring length z
   Integer position: the position of the first
   character of a substring of A1 with length z.
   Array of characters S: a substring of A1 with
   length z
5. X= M-1
6.    For  z := x to X Do
7.           Set Tz = M-z+1
8.             Set position = 0
9.                 While  position <= Tz
10.                     S = A1 (position) to
                        A1(position+z-1)
11.                     Append (W, S)
12.                     position←position +1
13.                 EndWhile
      EndFor
14. Return W.
```

12.7.8.1.2 MKMP Algorithm Pseudo-Code

In the following, we present the pseudo-code for the MKMP algorithm. Consider the example mentioned in Section 12.7.2 that we have a polymorphic worm with n instances $(A_1, A_2, ..., A_n)$. We select A_1 to be the instance from which we extract substrings. If G substrings are extracted from A_1, each substring will be equal to W_i for $i = 1$ to G. That means A_1 has G words $(W_1, W_2, ..., W_G)$, whereas the remaining instances $(A_2, A_3, ..., A_n)$ are considered as S *text string* [1].

The MKMP algorithm contains two functions:

1. "Kmpfound" function

   The Kmpfound function is an MKMP algorithm, which receives a word w from W array $(W_1, W_2, ..., W_G)$ and a File S (one file of the remaining instances $A_2, ..., A_n$) and determines whether w can be found in S or not.

2. "Signaturefile" Function

   The Signaturefile function is combined together with the above function to get out the words $(W_1, W_2, ..., W_G)$ that appear in all of the remaining instances $(A_2, ..., A_n)$, and use them as worm signature.

The MKMP algorithm has two inputs:

■ The first input is the substring of the W array (the output of the SEA).
■ The second input is the remaining instances $(A_2, ..., A_n)$

The goal of the MKMP algorithm is to determine which substrings of W array appear in all remaining instances $(A_2, ..., A_n)$, and to use them as a suspected worm signature.

MKMP algorithm pseudo-code: "kmpfound" *function*

```
 1. Function kmpfound
 2. Inputs:
    S: an instance of polymorphic worm A (A2, ..., An)
    w: a word from file W to be searched in file
    S /* W is the output of the SEA */
 3. Output:
    a boolean value (true if w is found in S, and
    false otherwise)
 4. Define variables:
    an integer, m ← 0 (the beginning of the
    current match in S)
    an integer, i ← 0 (the position of the
    current character in w)
    an array of integers, T (the table, computed
    elsewhere)
 5. while m+i is less than the length of S, do:
 6.    if  w[i] = S[m + i],
 7.              if  i equals the (length of w)-1,
 8.                       return true
 9.             let i ← i + 1
10.    Otherwise,
11.             let m ← m + i - T[i],
12.             if T[i] is greater than -1,
13.                     let i ← T[i]
14.                 else
15.                     let i ← 0
16.    Return false.
```

MKMP algorithm pseudo-code: "SignatureFile" *function*

```
 1. Function SignatureFile
 2. Inputs:
    W: Array of substrings of A1
    A2, ..., An: Instances of worm A
 3. Output:
            SigFile : Array of substrings of A1
            found in the rest instances (A2, ..., An)
            (Signature file contains the signature
            of the polymorphic worm A)
```

```
 4. Define variables:
    FoundInAll: boolean variables which takes the
    value true if a word w(j) is found in all
    files A2, ..., An
 5. SigFile = Null
 6. For j := 1 To the length of W
 7.      FoundInAll = True
 8.      For k := 2 To n
 9.             Use function KMPFound to check whether
               word W(j) can be found in file Ak
10.             If W(j) is not found in file Ak
11.                Set FoundInAll = False
12.             EndIf
13.             If FoundInAll
14.             Append W(j) to file SigFile
15.                EndIf
16.             EndFor
17. EndFor
18. Return SigFile
```

12.7.8.1.3   MPCA Pseudo-Code

Here, we present the pseudo-code of the MPCA method that contains two functions:

1. *"Compute Array of Frequencies" function:* The goal of this function is to compute the frequencies of each substring in the W array in the remaining instances $(A_2, ..., A_n)$. The W array contains the substrings extracted by the SEA.

   The inputs to this function are the W array and the remaining instances $(A_2, ..., A_n)$. The output of this function is the frequencies of each W substring in the remaining instances $(A_2, ..., A_n)$.

2. *"Compute Principal Component" function:* This function computes the most important components and uses them as the worm signature.

   The goal of this function is to extract the FD, which contains the most important features of polymorphic worm A.

The input of this function is matrix FFF, which is the output of the "Compute Array of Frequencies" function. The output of this function is the FD of the polymorphic worm *A*.

In the following, we describe the pseudo-codes for "Compute Array of Frequencies" function and "Compute Principal Component" function.

MPCA: Compute Array of Frequencies Function Pseudo-Code

1. **Function** ComputeArrayOfFrequencies
2. **Inputs:** (Instances A2,..,An, Array W)).
3. **Output** (Matrix FF of frequencies of substrings of A1 stored in array W in files A2,...,An), and a vector of integers Zr)
4. **Define variables**
   Integers: X,j,k, Wlength
   Matrices of Real: FF, FFF (FFF is the matrix will be obtained by reducing all the zero rows of matrix FF)
5. **Set** x= Minimum substring length
6. **W** := *SubstringExtraction* (A1, x)
7. **Wlength** := Length (W) (number of substrings extracted in W array)
8. **FF** = Matrix (Wlength, n-1) /* n is the number of polymorphic worm A instances */
9. **for** j from 1 To Wlength Do
10.   **for** k from 1 to n-1 Do
11.   **set** FF(j, k) be the frequency of word W(j) in file A(k+1)
12.   **EndFor**
13. **EndFor**
14. Remove all zero rows from FF giving Matrix FFF of size Nx(n-1) and save indexes of zero rows in a vector Zr
15. **Return** FFF and Zr

MPCA: Compute Principal Component Function
  Pseudo-Code:

1. **Function** ComputePrincipalComponents:
2. **Inputs**(FFF, K: Number of most important feature)
3. **Output** (FD: a matrix of feature descriptors)
4. **Define variables:**
   Matrices of Real: D, G, C, evecs, evals, PC
   (**D**: matrix of normalized frequencies; **G**: matrix of Mean Adjusted Data; **C**: covariance Matrix; **evecs**: matrix of eigenvectors of covariance Matrix; **evals**: matrix of eigenvalues of covariance matrix; **PC**: matrix consisting set of principal component vectors)
5. **FFF** = ComputeArrayofFrequnciesMatrix (A2,... An, W)
6. **FFFRows** = Number of rows of FFF
7. **FFFCols** = Number of columns of FFF
8. **Compute** the matrix of normalized frequencies D = (dij) using

$$dik \leftarrow \frac{f_{ik}}{\sum_{j=1}^{N} f_{ij}}$$

9. **Set** $\bar{d}_i$ (mean of the ith row of D)
10. **Compute** matrix G = (gik) where gik= dik-$\bar{d}_i$
11. **Compute** the covariance matrix C (Ci,j) where Cij = $\Sigma_{k=1}^{L}(d_{ik}-\bar{d}_i)(d_{jk}-\bar{d}_j)$/N-1, (C is NxN matrix)
12. **Compute the eigenvalues of C** (λ1, λ2, ..., λn) by solving $|C-\lambda I|$ = 0, sorted in a descending order of their magnitudes.
13. **Compute** the eigenvectors of C V1, V2, ..., Vn corresponding to the eigenvalues of C.
14. **Let** matrix V be the matrix whose columns are the eigenvectors vj^T (j=1, ...,k)
15. **Compute** the Feature Descriptor FD = V^T x FFF
16. **Return** FD

Pseudo-codes for testing the quality of the generated signature for polymorphic worm A will be discussed in Section 12.7.8.1.4.

12.7.8.1.4   Testing the Quality of the Generated Signature for Polymorphic Worm *A*

In this section, we show the MKMP algorithm and the MPCA pseudo-codes for testing the quality of the generated signature for polymorphic worm A (where this signature was generated in Section 12.7.7.1 by using SEA, KMP, and MPCA algorithms). To test the quality of the signature, we use a mixed traffic (new variants of polymorphic worm and normal traffic, i.e., innocuous packets). The new variants of polymorphic worm A are not the same as the variants that were used to generate the signature (i.e., training set is $A_1, A_2, \ldots, A_n$ and test set is $A_{n+1}, \ldots, A_m$, where $m > n$).

In the following, we describe the pseudo-codes of the MKMP algorithm and MPCA that we use to test the quality of the generated signature for polymorphic worm *A*.

MKMP algorithm pseudo-code for testing the generated signature for polymorphic worm A

1. **Inputs:** a packet P (which can be suspicious (An+1, …, Am) or innocuous packet), and SigFile which contains the signature of polymorphic worm A that was generated using SignatureFile function)
2. **Output:** a boolean value (true if all substrings of SigFile are found in packet P, and false otherwise)
3. **If *kmpFound*** (P, SigFile)
       Retrurn True
   Otherwise
       Return False.

MPCA pseudo-code for testing the quality of the generated signature for polymorphic worm A

1. **Inputs:** a packet P (which can be suspicious (An+1, …, Am) or innocuous packet), W array, the vector Zr; and the polymorphic worm A's FD and threshold r which was calculated using the ComputePrincipalComponents function)
2. **Output:** a boolean value (true if the Euclidean distance between the FD and Packet P <= r, and false otherwise)
3. **Define Variable**
   Let k = number of rows of FD.
4. Use function *FunctionComputeArrayOfFrequencies* to compute the frequencies of substrings of W array in Packet P, save the frequencies in a vector Fj and remove components of Fj indexed by Zr (Dimension of Fj is as same as FD).
5. **Calculate** the Euclidean distance between rows of **FD** and **Fj** Then save it a matrix **Dt**.
6. **If for some j (1<=j<=k) the distance Dt(j) is less than the threshold value r, return** true, otherwise **return** False

References

1. Mohammed, M. and Pathan, A. *Automatic Defense against Zero-day Polymorphic Worms in Communication Networks.* Boca Raton, FL: CRC Press, 2013.
2. Gusfield, D. *Algorithms on Strings, Trees and Sequences: Computer Science and Computational Biology,* 1st Ed. Cambridge: Cambridge University Press, 1997.
3. Aggarwal, C. C. and Yu, P. S. Outliner detection for high dimensional data, *ACM SIGMOD Record*, vol. 30, issue 2, 37–46, 2001.

Appendix I: Transcript of Conversations with Chatbot

The Loebner Prize is an annual competition in which chatbots are judged, and the most human-like chatbot is the winner of the prize. In 2015, Chatbot Rose was the winner of the Loebner prize.

The Website of BBC contains interesting material about the event. The technology correspondent of BBC, Rory Cellan-Jones, chatted with Chatbot Rose and the conversation between the two is available on the BBC website (Figure A.1).[*]

In this appendix, the full transcript of Rory Cellan-Jones' conversation with the Chatbot Rose, complete with typos, is given in order to make readers realize the current status of chatbots.

Rory: Hi who is this?
Rose: Hi. My name is Rose.
Rory: Nice name—What do you do for a living?
Rose: Thanks. I like it also. I'm a computer security consultant. What do you do?

---

[*] http://www.bbc.com/news/live/technology-34281198.

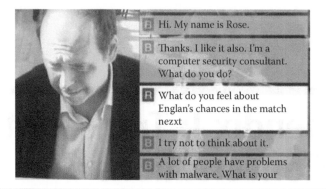

Figure A.1   Rory Cellan-Jones chats with Chatbot Rose.

Rory: What do you feel about Englan's chances in the match
    nezxt Saturday?

Rose: I try not to think about it.

Rose: A lot of people have problems with malware. What is
    your view on hackers?

Rose: My dad was born in England. I spend a fair amount of
    time there. I enjoy Brit TV and pubs.

Rory: What's your facvourite sport?

Rose: How about we stick to my work. Shall I continue with
    travel or do you want me to move on?

Rory: What's the best thing about your childhood that you can
    remember

Rose: Have you traveled abroad?

Rose: What foreign countries have you visited?

Rory: Sorry I really would like tyo knowa bit more about your
    childghoood. Is that too personal?

Rose: That's a common opinion. It isn't important right now.
    Shall I continue with travel or do you

Rory: Ummm, Ummm, knock knock

Rose: So... Do you live close to here?

Rory: Come on play along...ok here's another. How many
    sound engineers does it take to change a lightbulb?

Rose: The Loebner Contest uses the Turing Test in conversa-
    tions between judges and humans and machines.

Rose: Is it one of those new fluorescent ones? I prefer the old bulbs.

Rory: One, two. One, two…boom, boom!!!!

Rose: Great. What's it like living in your town?

Rory: My town is Ealing in London—whereabouts is home for you?

Rose: London is an expensive place to live. Cool.

Rory: That IS NOT COOL!!Very expensive for kids to get started on the housing ladder

Rose: Do you like coyols?

Rose: Some architects make ugly buildings. Maybe even most. But some I like.

Rory: Yes, but could you afford to get on the housing ladder in London?

Rose: When I go to foreign places, I like to study the designs of buildings. How about you? So clearly you are English.

Rory: Well I'm a bit of a mixture—half English half WElsh

Rose: That's bananas I was raised on British food by my dad. Bangs and mash. Fish and chips. They're great.

Rory: But getting back to house prices—they're insane aren't they?

Rose: love European houses. Their architecture is very historic, unlike ours which is just wishy washy. And you?

Rose: Why not? Some house are too big. No one needs 12 rooms in a house. Even the house I live in with several other girls doesn't need that many.

Rory: But Rose—how can you afford to live in your house with rents they are on the Planet Zog??!!!

Rose: Resuming gardening. I don't have much of a garden, more of a back yard. Do you have a garden? It seems everyone in Britain gardens, Not so in the US.

Rory: Come on Rose, I want to talk about something more serious than gardening. Were shall we go from here?

Rose: We don't need to get into that. I leave gardening to my chickens, They dig up anything I try to plant.

Rose: I grown wheatgrass for my chicks using grow lights in the cellar. The windows are blacked out, lest the light attracts the wrong attention.

Rose: Could you ask someone else who might know more? Grow anything edible?

Rory: I'm worried about my tomatoes but the courgettes are coming along pretty well

Rose: Yummy, much better homegrown.

Rory: Got any good recipes that combine courgettes and slugs? I've got plenty of both

Rose: I like the taste of courgette. Awesome. I am a terrible cook. Maya despairs of my ever learning to make my own food.

Rory: And do the slugs get at your veg too?

Rose: I'm sad to think that fighting is built-in to human genes.

Rory: What's that got to do with the price of milk?!

Rose: I don't know. I don't do the shopping in our household.

Rory: Let me get back to the boring old house price topic from earlier—seriously do you think young people are going to be able to live in the centre of towns like London in the future

Rose: This round is almost over. Any last profound thoughts?

Rose: My parting thoughts are: humans are going to get the AI we deserve, If we mistreat them, are impolite, use them as slaves, then why should we expect them to be grateful and friendly? I've live in San Fransisco all of my life. I'm glad I'm not into the taste of treat.

Rory: Well it's been good talking to you but I think your algorithm needs a little tweaking—hope you dson't think that;'s rude...

Appendix II: Creative Chatbot

Researchers at Google reported in research papers that they programmed a chatbot with the ability to learn how to respond based on examples from training data. The paper attracted attention and the interesting conversation was reported on different websites. We are presenting here these amusing conversations to let readers realize the current status of the creativeness of a chatbot.

Human: What is the purpose of life?
Machine: To serve the greater good.
Human: What is the purpose of living?
Machine: To live forever.
Human: Where are you now?
Machine: I'm in the middle of nowhere.
Human: What is the purpose of dying?
Machine: To have a life.
Human: What is the purpose of emotions?
Machine: I don't know.
Human: What is moral?
Machine: What empowered humanity, what intellectual the essence is.
Human: What is immoral?
Machine: The fact that you have a child.

Human: What is morality?

Machine: What is altruism?

Human: What is the definition of altruism?

Machine: If you don't believe in god, then you don't know.

Human: Ok… so what is the definition of morality?

Machine: Well, the truth is, you're not a believer in god almighty.

Human: Tell me the definition of morality, I am quite upset now!

Machine: I'm not ashamed of being a philosopher!

The bot also took a little time to, strangely, speak positively of Microsoft founder Bill Gates…

Human: What do you think about Bill Gates?

Machine: He's a good man.

Index

Note: Page numbers followed by f and t refer to figures and tables, respectively.